材料与医药化工专业英语

黄微雅 何冰晶 主 编
顾霞敏 徐倩倩 副主编

化学工业出版社

·北京·

内容简介

本书内容主要介绍了科技英语的特点和翻译技巧、材料制药化学类基本知识和专业文章阅读三大部分。专业文章阅读分为化学、化学工程与技术、材料科学、高分子材料和制药工程等专业，每个专业节选了适当难点和长短的该专业的英文原版文章。学生除了学习掌握本专业所节选的专业英语文章外，还可以将相近专业的专业英语作为选读内容，拓展知识面，提高专业英语的综合水平，培养学生的创新意识与创新能力。

本书可作为化学类、材料类和制药类相关专业，如应用化学、化学工程与技术、制药工程、无机非金属材料、材料工程、材料化学、高分子材料等学生的专业英语教材，也可以作为相关科技工作者提高专业英语知识的参考书。

图书在版编目（CIP）数据

材料与医药化工专业英语 / 黄微雅，何冰晶主编. —北京：化学工业出版社，2021.9（2025.2重印）
ISBN 978-7-122-39330-2

Ⅰ.①材… Ⅱ.①黄… ②何… Ⅲ.①材料科学-英语-高等学校-教材②制药工业-英语-高等学校-教材③化学工业-英语-高等学校-教材 Ⅳ.①TB3②TQ

中国版本图书馆 CIP 数据核字（2021）第 111933 号

责任编辑：张　蕾	文字编辑：毕梅芳　师明远
责任校对：王　静	装帧设计：刘丽华

出版发行：化学工业出版社（北京市东城区青年湖南街 13 号　邮政编码 100011）
印　　装：北京天宇星印刷厂
710mm×1000mm 1/16　印张 11　字数 223 千字　2025 年 2 月北京第 1 版第 6 次印刷

购书咨询：010-64518888　　　　　　　　　售后服务：010-64518899
网　　址：http://www.cip.com.cn

凡购买本书，如有缺损质量问题，本社销售中心负责调换。

定　　价：49.80元　　　　　　　　　　　　版权所有　违者必究

前言/Preface

专业英语作为专业知识传播和信息交流的媒介，已经成为各专业人士所必须掌握的一种国际性交流工具。材料与医药化工类专业英语是材料科学、制药、化学、化学工程等专业学生在完成两年大学基础英语和相关专业基础课程后可选修的一门课程，其主要目的是培养一批既具备高水平专业知识，又具备高水平外语交际能力的新型复合人才。

本书根据"由浅入深"的认知规律，结合材料与医药化工行业各个相关专业的专业英语特点进行编写。本教材主要分为三部分，第一部分为科技英语特点，主要包括材料与医药化工专业英语词汇特点及构词法、科技英语句子的特点和翻译技巧的介绍，帮助学生从基础英语逐渐过渡到专业英语的学习中。此外，增加了材料与医药化工专业论文写作技巧的介绍，培养学生在毕业论文中相关部分的英语写作能力。第二部分为材料与医药化工专业英语基础知识，主要包括元素周期律和元素周期表中常见元素的英文名称及其符号，有机和无机化合物的命名和有机化学、无机化学、物理化学、分析化学等相关的专业术语，便于学生扩大专业词汇量，为专业文章的阅读和理解奠定良好的基础。第三部分为专业文章的阅读，分为化学工程和工艺、材料科学和工程、高分子材料科学和工程和制药工程四个领域的专业文章，节选了部分原版英文教科书、专著及专业期刊中难度比较适中的文章，编排时对重要的专业词汇进行了注释和音标标注，课后附有练习题以巩固学习效果。使学生在掌握前面几项基本知识的基础上，进一步培养和进一步提高对专业文章的阅读能力。

本书内容较多，但由于时间仓促和编者水平有限，不妥之处希望读者提出宝贵意见，使本书在使用过程中不断改进和完善。

本书由黄微雅和何冰晶主编，顾霞敏和徐倩倩副主编。其中第一部分和第二部分由黄微雅和顾霞敏编写，第三部分由何冰晶和徐倩倩编写。此外，感谢台州学院专业英语教学组各位老师对本书的支持和建议。

编者
2021 年 1 月

Contents

Part I Scientific English Translating and Writing ········ 001

Unit 1 Brief Introduction to Scientific English ········ 001
Unit 2 Scientific English Translating Methods ········ 010
Unit 3 Writing a Scientific Paper in English ········ 026

Part II Basic English Knowledges for Pharmaceutical and Chemical Engineerings ········ 037

Unit 1 General Knowledge of Chemistry ········ 037
    Lesson 1 What is Chemistry? ········ 037
    Lesson 2 Chemical Bonding ········ 043
    Lesson 3 Working in Chemical Lab ········ 045
Unit 2 Nomenclature of Inorganic Compounds ········ 049
Unit 3 Nomenclature of Organic Compounds ········ 059

Part III Reading and Comprehension of Scientific Articles ········ 069

Unit 1 Chemical Technology and Engineering ········ 069
    Lesson 1 Chemical Industry ········ 069
    Lesson 2 Research and Development ········ 072
    Lesson 3 Typical Activities of Chemical Engineers ········ 075
    Lesson 4 Sources of Chemicals ········ 078
    Lesson 5 Basic Chemicals ········ 081
    Lesson 6 Ammonia ········ 083
    Lesson 7 What is Chemical Engineering? ········ 088

| Lesson 8 | Unit Operation in Chemical Engineering | 091 |

Unit 2 Materials Science and Engineering · 095
Lesson 1	Introduction to Materials Science and Engineering	095
Lesson 2	Solid Materials and Engineering Materials	097
Lesson 3	An Introduction to Metallic Materials	100
Lesson 4	Introduction to Ceramics	102
Lesson 5	Polymeric Composite Materials	107
Lesson 6	Materials and Technology	110
Lesson 7	Introduction to Nanoscale Materials	111
Lesson 8	Applications of Nanomaterials	114

Unit 3 Polymer Science and Engineering · 117
Lesson 1	Polymer	117
Lesson 2	What Are Polymers?	121
Lesson 3	Polymer Synthesis	125
Lesson 4	Chain Polymerization and Bulk Polymerization	129
Lesson 5	Molecular Weight and Its Distributions of Polymers	131
Lesson 6	Structure and Properties of Polymers	133
Lesson 7	Functional Polymers	136
Lesson 8	Preparations of Amino Resins in Laboratory	138
Lesson 9	Processing and Fabrication of Thermoplastics	139

Unit 4 Pharmaceuticals Engineering · 143
Lesson 1	Production of Drugs	143
Lesson 2	Isolation of Caffeine from Tea	149
Lesson 3	The Chemistry of Insulin	153
Lesson 4	Tablets (The Pharmaceutical Tablets Dosage Form)	156
Lesson 5	Sterile Products	159
Lesson 6	Natural Products	162
Lesson 7	Discovery of Sulfa Drugs	165

References · 168

Part I

Scientific English Translating and Writing

Unit 1 Brief Introduction to Scientific English

Along with the rapid development of science, technology, culture and education, more and more scientific English books and technical documents dealing with the achievements in all fields have been coming out. Scientific English, as a specific technical language, is popular over the world and has its important role. However, people have been facing lots of difficulties in reading the works written in scientific English, since scientific English has its own characteristics in using words, phrases and language structures. The difficulties should be attributed to the different expressions between ordinary English and scientific English. Understanding and studying these characteristics will help us to better master the scientific knowledge.

Ordinary English is an essential aspect of the modern English. It often uses simple words or phrases to express complicated meanings. Eventhough the readers do not understand the exact meanings of some new words or technical jargons in certain fields, they can guess the meanings according to the context and grammatical constructions. In one word, the main characteristic of ordinary English is simple, popular and practical to life. On the other hand, scientific English has its particular idiomatic expressions other than those in ordinary English. Both its type of writing and grammatical structure, may be different, too.

Especially, ordinary English and scientific English have different expressions in using words, voices, and idioms. Now some examples about the different expressions between the two forms of English are given below. Sentences in (a) are written in ordinary English, and those in (b) in scientific English, both having the same contents.

(1) The difference in using words or phrases

Different words of the same meaning are chosen for different forms, and the grammatical structure may also different.

(a) As the rubber industry grew, people needed more and more rubber.

(b) As the rubber industry developed, more and more rubber was required.

In the above sentences, "developed" has the same meaning as "grew", "needed" and "was required" are of the same meaning but of different voices. These examples show that a synonym and a passive voice are used in the scientific English sentences.

(2) The different expression in using voice

It is a remarkable characteristic style of scientific English to use the passive voice much more frequently than ordinary English. It is probable that in scientific research reports at least one-third of the predicate verbs are in passive voice. Nevertheless, active voice is most frequently used in ordinary English.

(a) When you boil a kettle of water, you can see steam coming out of the spout.

(b) When a kettle of water is boiled, steam can be seen coming out of the spout.

In (a), the personal pronoun used as the subject does not give the readers any new information. In (b), "a kettle of water", the subject, can attract the reader's attention.

一、科技英语的特点

1. 科技英语的特点

科技文体崇尚严谨周密，概念准确，逻辑性强，行文简练，重点突出，句式严整，少有变化，常用前置性陈述，即在句中将主要信息尽量前置，通过主语传递主要信息。科技文章文体的特点是清晰、准确、精练、严密。

2. 科技英语和普通英语的比较

普通英语（即基础英语、日常英语）涉及全部语言技巧，侧重于口语与听力能力的培养，包括英语语法及构词，朗读是学习普通英语的基本功之一。普通英语没有太多的涉及各个方面的专业术语，但是普通英语是专业英语的基石，普通英语掌握不好会很大程度上影响专业英语的学习。只有在学好普通英语的基础上，才能掌握科技英语。Table 1 为日常英语与科技英语的比较。

科技英语作为英语中的一种语体形式，它不仅与普通英语（即基础英语、日常英语）具有共性，又有自己显著的特性：

科技词汇的意义具有单义性、准确性和稳定性；

科技英语句子长、结构繁、朗读难；

科技英语中被动语态用得较多，翻译时必须根据汉英两种语言的习惯用法加以适当处理；

科技英语中虚拟语气经常出现。

Table 1　Ordinary English VS Scientific English

Ordinary English	Scientific English
通俗化：常用词汇用得多	专业化：专业术语用得多
多义性：一词多义，使用范围广	单义性：词义相对单一，因专业而不同
人称化：人称丰富，形式多样	物称化：多用物称，以示客观
多时性：描述生活，时态多样	现时性：叙述事实多用现在时
主动性：句子倾向于用主动态	被动性：句子倾向于用被动态
简单性：单句、散句用得多	复杂性：复杂单句、复合句用得多
口语化：口语用得多，随意灵活	书面化：长句用得多，书卷气浓

二、科技英语词汇的特点

随着科技的飞速发展，国际交流和合作日益频繁，社会对各类职业人才的要求越来越高，要求他们不仅具有扎实的专业知识，而且具有一定的专业英语水平，能用英语进行专业信息交流。其中，医药化工行业的迅猛发展令人瞩目，而且该行业涉及化学、化工、制药、材料等众多相关专业，故其专业词汇纷繁复杂，数量巨大，往往难以掌握，但从构词角度不难发现这些词汇中很多都是由我们所熟悉的单词通过加缀复合或缩略而形成的。例如，在英汉技术词典中，以 multi-构成的词有 400 个以上，以 micro-构成的词有 300 个以上。因此，很好地掌握专业词汇的构词特点，可为学习者提供一个有效的学习和记忆医药化工专业英语词汇的科学方法和快速扩大词汇量的捷径。专业英语词汇有以下几个特点。

1. 专业术语

专业英语中的大量专业词汇，词形长、难念，不像文学词汇词形悦目、易于上口。但技术词汇词义专一，能用来表达确切的含义。例如：

hydrolyze——水解；alkane——烷烃；evaporation——蒸发；aluminum——Al（铝）；oxidation——氧化；methylamine——甲胺；halogen——卤素；nitroglycerin——硝酸甘油

2. 半专业词汇

(1) 常用词汇专业化

单词	普通词义	科技词义	单词	普通词义	科技词义
solution	解决方案	溶液	support	支持	支架
order	次序、命令	级数	function	功能	函数
power	力量	功率	flash	闪耀	闪蒸

续表

单词	普通词义	科技词义	单词	普通词义	科技词义
die	死	模具	bird	鸟	飞机、火箭
monitor	班长	监视器	monkey	猴子	活动扳手、起重机
cock	公鸡	旋塞、吊车	fox	狐狸	绳索

(2) 同一词语词义多专业化
例如：

单词	transmission				
专业	电气工程	无线电	机械学	物理学	医学
含义	输送	发射、播送	传动、变速	透射	遗传

又如：

单词	phase				
专业	土壤学	电工学	数学	物理学	军事学
含义	分段	相位	位相	相	战斗阶段

3. 前缀和后缀（prefix and suffix）出现频繁

许多专业词汇为了使自身的词义得以扩展，往往通过加前缀和后缀的方式获得很强的派生能力。前后缀出现频繁是医药化工类专业词汇的构词特点之一。

例如：词缀 hydro-表示氢的、含氢的，将该词缀加在单词 carbon（碳）之前则派生出新的专业词汇 hydrocarbon，该专业词汇的含义则为碳氢化合物，即烃类。又如：hydroxide 氢氧化物；hydroboron 氢化硼；hydrochloric acid 盐酸；hydrocrack 氢化裂解；hydrofine 氢化提纯；hydrotreat 氢化处理；等等。

4. 较多使用复合词

医药化工类专业术语中有不少是由两个或两个以上的词复合构成的，这类词在构词法中被称为复合词，其含义往往与组合成复合词的各单词的含义相关。复合词组成面广，多数以短划线-连接单词构成，例如：

by-product 副产物；rust-resistance 防锈；scale-up 放大；pilot-plant 中试工厂；cross-link 交联；silica-gel 硅胶；surface-active agent 表面活性剂；等等。

或者采用分写式，无连字符，例如：

atomic weight 原子量；heat exchanger 热交换器；flow sheet 工艺流程图；solubility product 溶度积；free radical 自由基；stationary phase 固定相；electrical conductivity 导电

性；functional material 功能材料；等等。

5. 广泛采用缩略词

缩略词的目的是为了以简洁的方式表达高度浓缩的内容和概念，这正是医药化工类专业词汇的又一重要特征。例如：

USP　United States pharmacopeia　美国药典；
GMP　good manufacturing practices　良好操作规范；
CRE　chemical reaction engineering　化学反应工程；
HPLC　high performance liquid chromatography　高效液相色谱；
PVC　poly (vinyl chloride)　聚氯乙烯。

三、专业英语词汇的构词法

1. 转化词

n.→v.：water 水→浇水　　　charge 电荷→充电　　　yield 产率→生成
a.→v.：dry 干的→烘干

2. 复合词（compounding）

复合词又称合成词，是由两个或两个以上的词合成一个新词。

副词+过去分词	well-known	著名的
名词+名词	carbon steel	碳钢
	rust-resistance	防锈
名词+过去分词	computer-oriented	面向计算机的
介词+名词	by-product	副产物
动词+副词	makeup	化妆品
	check-up	检查
形容词+名词	atomic weight	原子量
	periodic table	周期表
动词+代词+副词	pick-me-up	兴奋剂
副词+介词+名词	out-of-door	户外

3. 派生词（derivation）

派生词是指利用词的前缀与后缀作为词素构成新词。一般字典上查不到。
① 前缀一般要有固定的意义
◆ mono-（单）　　　　　monocrystal/monophase/monoatomic

- di-（双） dipole/carbon dioxide
- bi-（双） bicycle/biphase/binary compounds
- poly-（聚） polymer/polyethylene/polyatomic
- micro-（微） microcomputer/micrometer/microfiltration
- semi-（半） semiconductor/semiabstract/semi-official
- anti-（抗，反） antisymmetric/antistatic

② 后缀一般表示词类
- -less noiseless/stainless steel
- -phone earphone/telephone/ cellphone
- -ics electronics/physics
- -ane methane/ethane/propane/butane...
- -ene ethene/propene/butene...
- -yne ethyne/propyne/butyne...

4. 混成词（blending）

混成词是取两个词中在拼写上或读音上比较合适的部分组成一个新词。通常是将两个单词的前部拼接、前后拼接或者将一个单词前部与另一个单词拼接。

- smog=smoke+fog
- electrochemical=electric+chemical
- aramid=aromatic+amide
- redox=reduction+oxidation
- escalift=escalator+lift
- pictogram=picture+telegram
- multivider=multiply+divider

5. 缩略词（shortening）

缩略词是将较长的英语单词取其首部或者主干，构成与原词同义的短单词，或者将组成词汇短语的各个单词的首字母拼接成一个大写字母的字符串。

(1) 压缩和省略

将某些太长、拼写难记、使用频繁的单词压缩成一个短小的单词，或取其头部、或取其关键音节。

例如：flu=influenza 流感　　lab=laboratory 实验室　　math=mathematics 数学

(2) 缩写

将某些词组和单词中每个实义单词的第一个字母或者首部几个字母重新组合，组成一个新的词汇，作为专用词汇使用。

① 通常以小写字母出现，并作为常规单词。

radar=radio detecting and ranging　　　　　　　　雷达=无线电探测与定位

laser=light amplification by stimulated emission of radiation　激光=受激光发射光放大器
modem=modulator demodulator　　　　　　　　　　　　　调制解调器
② 以大写字母出现，具有主体发音音节。
BASIC=beginner's all-purpose symbolic instruction code　初学者通用符号指令代码
TOEFL=test of English as a foreign language　　　　　　　作为外语的英语考试（托福）
CET=college English test　　　　　　　　　　　　　　　大学英语考试
SOS=save our souls　　　　　　　　　　　　　　　　　国际通用呼救信号
③ 以大写字母出现，没有读音音节，仅作为字母缩写（见科技英语特点）。
FDA=Food and Drug Administration　　　　　　　　　　（美国）食品及药物管理局
DNA=deoxyribonucleic acid　　　　　　　　　　　　　　脱氧核糖核酸
PPM=parts per million　　　　　　　　　　　　　　　　百万分率

四、专业词汇的翻译

1. 音译

① 计量单位名称（一般采用音译）：hertz, newton, pascal, coulomb（库仑）；

② 新型材料名称（一般采用音译）：nylon, vaseline, morphine, tannin, caffeine, heroin；

③ 由人名构成的术语，一般采用音、意混合译法：Faraday constant, Avogadro constant, Lenz's law, Joule's law, Joule-Thomson effect（焦耳-汤姆孙效应），Einstein equation, Bunsen flame（本生灯）。

2. 根据专业或学科选择词义

同一词在不同场合往往有不同含义，应根据上下文，即根据专业特点和具体语境来确定其确切含义。例如：power

① A car needs a lot of *power* to go fast. 汽车高速行驶需要很大的<u>动力</u>。

② *Power* can be transmitted a very long distance. <u>电力</u>可以传送到很远的地方。

③ The fourth *power* of three is eighty-one. 3 的 4 <u>次方</u>是 81。

④ The combining *power* of one element in the compound must equal to the combining *power* of the other element. 化合物中一种元素的<u>化合价</u>必须等于另一元素的<u>化合价</u>。

⑤ Energy is the *power* to do work. 此句 power 表示"能力"。

⑥ *Power* is the rate of doing work. 此句 power 表示"功率"。

⑦ Basically, all *power* is with the people. 此句 power 表示"权利"。

⑧ Political *power* grows out of the barrel of a gun. 此句 power 表示"政权"。

⑨ Knowledge is *power*. 此句 power 表示"力量"。

⑩ The surfaces of ceramics must be checked by examining it under a 20 *power* binocular microscope. 此句 power 表示"倍（数）"。

其他词汇：

① This medicine *acts* well on the heart. 这种药对心脏的**疗效**很好。

② *Work* must be done in setting a body in motion. 使一个物体运动时，必须做**功**。

③ Even though *bearings* are usually lubricated, there is friction and some wear. 即使经常给**轴承**润滑，还存在摩擦与一些磨损。

3. 根据上下文选择词义

英语的词义对上下文的依赖比较大，独立性比较小。一个词的意义往往取决于上下文的意思。

① It is impossible to predict in detail the shape and *mechanism* of the robot salve. It might carry its computer and response *mechanism* around with it and also its source of power.

② Metallic iron contents were determined by electrochemical *solution* of the iron with copper nitrate *solution*.

4. 根据词的搭配来选择词义

large current 强电流　　large capacity 高容量　　large loads 重载
large-screen TV 宽屏电视　large growing 生长快的　a large amount of money 一大笔钱

5. 词义的引申

① You found that, in two experiments, *hardness and greenness in apples went together with sourness*.

② These vacuum tubes *will see use* in the output stage.

③ Materials science—once the least *sexytechnology*—is *bursting with* new, practical discoveries.

6. 词义的转换

① Despite all the *improvements*, rubber still has a number of limitations. 尽管**改进**了很多，但合成橡胶仍有一些缺陷。（n.→v.）

② Boiling point *is defined* as the temperature at which the vapor pressure is equal to that of the atmosphere. 沸点的**定义**就是气压等于大气压时的温度。（v.→n.）

Exercises

1. Try to translate the following into Chinese according to word building

bio-, biomaterial/biotechnology/biodegradable/biomedicine/biocatalyst/biosynthesis
nano-, nanochemistry/nanostructure/nanotechnology/nanotube/nanosecond
anti-, antibody/anticancer/antioxidant/antifoam/antitumor

2. Try to translate the following word ("good") into Chinese

good knife   good conductor   good English

good money   good soil   good oil

Milk is good food for children.

It is no good heating the material to such a temperature.

The workers gave the machine a good checking.

making good use of the sun

Unit 2　Scientific English Translating Methods

Difficulty in understanding professional expressions

　　Scientific terms refer to the expressions in science referring to certain professional concepts. Scientific terms are professional with many meanings; so they should be dealt with according to professional content in translation process, otherwise, great errors may happen with a little carelessness. Here is an example: the term "cassette", except for other meanings, as audio tape, has two meanings, "packing tape box" and "case style tape", and the meaning should be decided according to the context. For example, in the specifications of imported radio recorder, there are two subtitles: "checking the cassette" and "to insert cassette", in which the meanings are different; the first cassette means the tape and the second means the box.

　　There are many differences and changes in the expressions of Chinese and English, and the translation should not only correspond to the original text in expressions, but also be proper in the place and times of expressions, to be smooth and exact. Professional scientific English translation should be able to be faced with numerous professional environment, where many words have more than one meaning and present in abbreviations, and therefore, it is hard to be referred in ordinary dictionary; so it is very important to translate these expressions exactly. However, some words seem ordinary in form, but they have different meanings in particular professional environment, which requires translators to know deeper about professional environment.

Grasp the exactness of vocabulary context

　　The meaning of certain word should be defined before understanding its class. Many words have total different meanings in ordinary English and scientific English, as well as in different majors. In the translating process, meanings of words should be defined according to the content of majors. During translating process, sometimes it is hard to find proper meaning in English dictionary, in which case, if one translates word-for-word randomly, the translated articles would be stiff and obscure and cannot express the original meaning exactly, which may cause misunderstanding. Especially when the sentence is too long or there are too many subordinated clauses, translators should understand the lingual expressing structure clearly and analyze the logical relationship deeply, with many translating methods to transfer language expressing forms, so as to avoid word-for-word translation.

　　In scientific English, the subject part of a sentence always contain large amount of information and has a complex structure.

For example:

a. "x" is slightly over (under) 3 cm long.
b. "x" has an approximate length of 3 cm.
c. the approximate length of "x" is 3 cm.

In these three sentences, scientists would accept the third sentence which proposed information.

一、科技英语句子的特点概述

科技文体崇尚严谨、周密，概念准确，逻辑性强，行文简练，重点突出，句式严整。因此，句子组成方法有其自身的特点。

1. 大量使用名词、名词性词组或短语和名词化结构（more nouns、noun phrases and nominalizations）

科技文体是以事实为基础论述客观事物。作者在遣词造句中要求客观地表达事物特性，避免主观意识，强调存在的事实，而非某一具体行为，名词、名词性词组正是表物的词汇。所以，在普通英语中用动词等表示的内容，科技英语惯用名词来表达，把原来的施动含义蕴藏在结构中。由于大量使用名词、名词性词组，也就必然要多用介词，从而构成较多的名词短语和名词化结构。它们以短语形式来表达，相当于一个句子所要表示的内容。

例如：Scientific exploration, the search for knowledge has given man the practice results of being able to shield himself from the calamities of nature and the calamities imposed by other man. 科学的探索，知识的追求，使人类获得了避免天灾人祸的实力。

这个例句中包含十个名词（包括动名词），其中有五个名词性词组（短语），却只有一个谓语动词和一个动词不定式。多用名词和少用动词，原是英语的特点，在科技英语中更为突出。例句中 the search for knowledge 是名词化结构。所谓名词化结构，是指以动词名词化的词为中心词，和其有内在逻辑关系的修饰语构成一个名词短语。它实际上是一个句子的压缩。动词名词化词包括行为动词（与动词同根或由动词派生）、名词性动名词（不具备动词特性）和动名词（兼有动词特性）三种，为了叙述方便，简称"动作性名词"（或行为名词）。

2. 被动结构比较普遍（extensive use of the passive voice）

科技英语在句法上比较突出的特点之一是被动语态使用多，几乎占了全部谓语动词的三分之一以上。C.C.巴伯统计表明，在三份不同的材料中，一共出现了1475个动词，其中28%为被动形式，72%为主动形式。被动结构之所以受到科技作者如此的青睐，主要是因为被动结构具有突出主题、引入主题、转换主题、凸显焦点和语段衔接与连贯五种语篇功能，具体来说有以下几个因素。

① 科技作者在探讨事物的发展过程和阐述科学原理时，往往着眼于演绎论证的结果，而不太关注动作的实行者。这就需要使事物、过程和结果处于句子的中心地位，而被动结构正好能突出要论证和说明的对象。

② 科技作者对现象的描述、规律的论证、事理的分析和事物相互关系的推理，力求尽量客观，而客观的表达与感情色彩有冲突，被动结构正好可以避开人的主观因素。

③ 科技文章崇尚准确、严谨和精练，而被动结构在很多情况下可以使句子更加紧凑和简短。

④ 便于扩展句子，而不至把句子弄得头重脚轻，保持句子平衡，同时也符合主要信息前置的原则，这也是科技英语的描述功能、分类功能、定义功能和指令功能（间接指令）所决定的。例如：

① When the radiant energy of the sun falls on the earth, it is changed into heat energy, and the earth is warmed.

太阳的辐射能到达地球后就转化为热能，从而使大地暖和起来。

② The Harry Diamond Laboratories performed early advanced development of the Arming Safety Device (ASD) for the Navy's 5-in guided projectile. The early advanced development was performed in two phrases.

哈里-代蒙德实验室对美海军 5 英寸制导炮弹的解除保险装置（ASD）进行了预研。预研工作分两个阶段进行。

3．广泛使用非谓语动词

科技英语中，动词的非谓语形式（分词、动名词、动词不定式及其复合结构）广泛使用，尤其是分词短语用作后置定语的现象更是层见叠出。形成这种情况的原因主要有下面几点。

① 因为非谓语动词没有人称和数的区别，也无情态动词，它们有两个基本特点：一是阐述概念的客观性强；二是无主语，常作为句法上的压缩手段，所以使用频率较高。

② 在科技作品中，人们往往要说明各个事物之间的关系、事物的位置和状态变化，如机器、产品、原料等的运动、来源、型式、加工手段、工艺流程和操作方法，这些都要求叙述严谨、准确。动词的谓语形式容易满足这些要求，而且能用扩展的成分对所修饰的词进行严格说明和限定（每一个分词定语都能代替一个从句）。

③ 科技作者为了完整、准确地表达某一概念和事物，常需对某些词句进行多方面的修饰和限定。为了使很长的句子匀称，避免复杂结构并省略动词时态的配合，使句子既不累赘又语意明确，往往采用非谓语动词这种语法手段。

例如：

① China was the first country to invent rockets. 中国是第一个发明火箭的国家。

② Astronauts performing tasks abroad space shuttle get help from robot arm. 宇航员在航天飞机外面执行任务时可借助于机械手。

③ The volume of fuel oil extracted from the liquid produced increases substantially. 从所产生的液体中提取的燃料油的数量大大增加。

④ Cement, wood and steel are the most widely used building materials. 水泥、木材和钢材是最广泛使用的建筑材料。

⑤ A safety valve is provided in order to allow excess pressure to escape. 设置有安全阀，以降低过高的压力。

4．无人称句使用多（more impersonal sentences）

John Swades 认为，科技英语中，有五种主语形式：祈使结构、被动结构、第一人称单数、第一人称复数、第二人称复数。他强调第一、第二种结构比较妥当和常用。特别是在过程描述、功能描述、物理描述、因果描述等描述功能、定义功能、分类功能和指令功能中采用，它使科技英语显得更加客观。因此，科技文体一个显著的特点就是无人称。可以说，绝大多数科技文章（scientific articles）很少使用有人称的句子。这主要是由于科技文章所描述和讨论的是科学发现或科技事实。尽管科技活动系人类所为，但科技文章所报道的主要是科技结果或自然规律，而不是报告科技结果或自然规律是由谁发现或完成的，因此科技文章往往没有人称。科技文章将主要信息前置，放在主语部分。这也是广泛使用被动态的主要原因。试观察并比较下面两段短文的主语。

We can store electrical energy in two metal plates separated by an insulating medium. We call such a device a capacitor, or a condenser, and its ability to store electrical energy is called capacitance. It is measured in farads.

Electrical energy can be stored in two metal plates separated by an insulating medium. Such a device is called capacitor or condenser. Its ability to store electrical energy is capacitance. It is measured in farads.

译文：电能可储存在由一绝缘介质隔开的两块金属极板内。这样的装置称之为电容器，其储存电能的能力称为电容。电容的测量单位是法拉。

后一段短文中各句的主语分别为：Electrical energy；Such a device；Its ability to store electrical energy；It (capacitance)。它们都包含了较多的信息，并且处于句首的位置，非常醒目。四个主语完全不同，避免了单调重复，保证前后连贯，自然流畅。足见被动结构具有简洁客观的特点。

二、科技英语翻译的要求

The three-character principle proposed by Yan Fu in his translation of Evolution, Ethics and Other Essays are faithfulness, expressiveness, and elegance.

翻译的要求：信、达、雅。

三、科技英语常用句型及翻译

1. 被动语态的常用原因

(1) 不必或无法说出主动者

For a long time aluminum has been thought as an effective material for preventing metal corrosion. 长期以来，铝被当作一种有效防止金属腐蚀的材料。

The book has already been translated into many languages.

(2) 强调行为对象，而非行为者，将行为对象作为句子的主语

Three machines can be controlled by a single operation. 三台机器能由一个操作者操纵。

The work must be finished at once.

(3) 更好地联系上下文

They are going to build a library here next year. It is going to be build beside the classroom building.

2. 被动语句的翻译方法

(1) 仍译为被动句

最常见的是在谓语前加上助词"被"，也可使用"受到、遭到、得到、叫、称、让、给、加以、为……所"等句式。偶尔也可不加任何词直接译出。

We remember how air can be made into a liquid. If the liquid is warmed again, it "boiled" and turns back into a gas. 我们记得是如何把空气制成液体的。如果给这种液体重新加热，它"沸腾汽化"，就又还原成气体。

Every thing is built up of atoms.

The process by which energy is changed from one form into another is called the transformation of energy.

(2) 译为主动句，符合汉语的习惯

① 译成无主语的主动句。

Measures have been taken to diminish air pollution. 已经采取了一些措施来减少空气污染。

If the product is a new compound, the structure must be proved independently.

Quality must be guaranteed first.

② 加上不确定主语的主动句。

Salt is known to have a very strong corroding effect on metals. <u>大家</u>知道，盐对金属有很强的腐蚀作用。

All bodies are known to possess weight.

If one or more electrons are removed, the atom is said to be positively charged.

③ 由 by 短语中的动作发出者作为主语。

Heat and light are given off by the chemical reaction. <u>这种化学反应</u>能发出热和光。

The planets are held together by attractive forces.

Every day, new compounds are produced in the laboratory by the synthetic organic chemist.

(3) 译成判断句

Currently most solar cells are made from crystals of high-purity silicon. 目前，绝大多数太阳能电池<u>是</u>用高纯度的晶体硅制成<u>的</u>。

Large quantities of oil are refined locally at Abadan, Iran, in Bahrain and elsewhere.

(4) 译成因果句

The crops were badly damaged by a flood. 农作物因水灾而毁坏。

The multiplicity of carbon compounds is explained in the exceptional ability of carbon atoms to combine with one another. 碳化合物之所以为数众多，是因为碳原子相互结合的能力特别强。

The existence of organic chemistry is further justified by the importance of carbon compounds or mixtures in our lives.

四、各种从句的翻译

大量使用复合句是科技英语的重要特征之一，一个复合句可能由主句和多个从句组成。理解和掌握各种从句的译法对理解和翻译科技英语是非常重要的。

1. 主语从句

(1) 基本句型：连词（连接代词、连接副词）+主语+谓语+谓语

<u>Why dinosaurs became extinct</u> remained a mystery for a long time. 恐龙为什么绝迹长期以来是个谜。

<u>Whether there is life on the Mars</u> is a disputable topic.

(2) 用 what 引起的主语从句

<u>What we call weight</u> is really the gravitational pull on an object. 我们所称的重量实际上是对某一物体的地心引力。

<u>What cannot be seen with the eye</u> can be easily found with a radar instrument.

(3) 用 it 做主语的复合句

It is important <u>that the metal for making aeroplanes should resist corrosion.</u>

2. 宾语从句

(1) 带有宾语从句的复合句：一般按原句顺序翻译

A product must be carefully tested to determine <u>if it will perform its job properly and reliably</u>. 产品必须仔细检查，以确定是否合格可靠。

Newton believed <u>that gravitation is everywhere—that is, it is universal.</u>

(2) 用 what 引起的宾语从句

常有"the thing which"的含义，译为"所……的东西（或事物）"：

The computer can only do <u>what they are told to do</u>. 计算机只能做人们所要它做的事情。

When the atoms of different elements combine or react chemically, they form <u>what are called "molecules"</u>.

(3) 介词所带的宾语从句

除动词外，介词也可带有宾语从句，这时从句不能用 that 引导，而是用连接代词或连接副词引导，翻译时要把宾语从句处理成一个短语。

Energy is rather difficult to define, but most of us have a concept of <u>what it is</u>. 给能量下定义是相当不容易的，但我们大多数人对这一概念都有所了解。

The value of a machine depends on <u>how much work it can do per hour, or per second</u>.

(4) 复合宾语中的宾语从句

Time and again Newton made it clear <u>that motion is relative</u>.

3. 表语从句

(1) 表语从句一般译法

One of the most important facts about light is <u>that it travels in straight line</u>.

(2) 由 what 引起的表语从句

Oil coming from the ground is <u>what we call crude oil</u>. 从地下喷出的石油就是我们所说的原油。

Carbon is <u>what makes up a major ingredient in coal, wood, gasoline, oil, or any other fuel</u>.

4. 同位语从句

① 把同位语从句译在其先行词之前，两者之间用"这一"或"的"连接。

The question <u>whether there is life on Mars</u> is still disputable. 火星是否有生命这一问题仍在争论之中。

The distilling unit separates oil into different products by taking advantage of the fact <u>that the hydrocarbons boils at different temperatures</u>.

② 按照原文顺序翻译，在先行词前加"这一"等词，其后加冒号或破折号，然后译出同位语从句，这种方法只能用在同位语从句及其先行词不在句首的复合句中。

We are thus led to the conclusion <u>that friction is not always something undesirable</u>.

5. 定语从句

(1) 限制性定语从句

① 限制性定语从句是句子中不可缺少的部分，如果除去它整个句子就可能失去意义。在翻译时通常省去关系代词而把定语从句放在所修饰的名词之前。

Anything that takes up space and has weight is matter. 任何占据空间、具有重量的东西都是物质。

② 在限制性定语从句中，做宾语的关系代词有时可省去。

The heat energy we get from the sun may be changed into other forms of energy. 我们从太阳得到的热能可以变成其他形式的能量。

(2) 非限制性定语从句

① 译为并列句，关系代词译作它所替的词，前面加上"这个""这些"，也可以不重复所替代的词，而用"它们"等表示。例如：

The most familiar ferrous alloy is steel, which contains iron and a very small proportion of carbon.

② 有时主句和从句之间有转折或连续关系，在翻译时除照上述方法外，还可以加上"而""可是""但"等连词，例如：

An atom consists of a nucleus, which contains protons and neutrons, surrounded by electrons in one or more shells.

③ 有时定语从句含有原因、让步、条件等意义，这时可以将其译成相应的状语，例如：

Computers, which have many advantages, cannot take the place of man. They can only do what they are told to do. 计算机虽然有许多优点，但是不能代替人。它只能做人们要它做的事情。

The water still cooled off quickly at night through the glass, which was exposed to the cold night air.

(3) 关系代词 which 前有介词的译法

① 介词是从句中的某个动词或短语所要求，翻译时把介词作为短语的一部分，不专门译出。

The degree to which they impede the flow of current is called resistance. 阻碍电的流动的程度叫作电阻。

Electricity is a subject about which many books have been written.

有的介词用于其本身含义，表示地点、时间、原因等，应按其含义译出。

A machine is a mechanical contrivance by means of which a force is used to overcome a resistance. 机器是一种借助它施加力以克服阻力的机械装置。

② 在 which 前用 of 表示"部分"概念时，关系代词常译为"其中"。

An acid was originally defined as a compound containing hydrogen, some or all of which could be replaced by a metal to form a new compound known as a salt.

(4) 关系代词 whose 在定语从句中用作定语的译法

关系代词 whose 在定语从句中用作定语可译为"它的，它们的，其"。

All matter comprises atoms and molecules, whose spacing depends on the strength of the intermolecular forces. 所有物质都由原子和分子组成，其间距取决于分子间诸力的大小。

(5) 由 where 引导的定语从句

Naturally, solar water heaters work best in climates <u>where there is a lot of sunshine and where the temperatures are quite warm</u>. 当然，在阳光充足、温暖和煦的天气，太阳能热水器运行得最好。

See Table 4, <u>where all metals are listed in the order of their specific weight.</u>

(6) 由 which 引导的特种定语从句

All matter has mass, <u>which is known to all</u>. 一切物质都有质量，这是众所周知的。

The specific heat of water is 1, <u>which means that it requires one calorie of heat to raise one gram of water one degree.</u>

(7) 用 as 引导的特种定语从句

<u>As we know</u>, the conductivity of semi-conductors increase with the increase of temperature. 正如我们所知，半导体的导电性随着温度的升高而增大。

Power equals the square of the current multiplied by the resistance, <u>as (is) indicated in the formula $P=I^2R$.</u>

6. 状语从句

(1) 时间状语从句

① 英语中时间状语从句可以放在句首、句中或句末，译成汉语时时间状语从句通常放在主句之前。

<u>When we speak of a force being responsible for motion</u>, it is not enough for us to tell only its magnitude. 当我们谈到一个力引起运动时，只谈力的大小是不够的。

An electromagnet is only a magnet <u>while a current is flowing</u>. 只有电流通过时，电磁铁才有磁性。

② 用 until 或 till 连接的从句一般在翻译时放在主句之后，译为"直到……为止"。

A theory in physics does not become a law <u>until it has been tested and shown to apply in many previously untested areas.</u> 物理学中的理论要经过检验，并证明可以在许多以前未试验过的领域中应用，才能变成定律。

The development stage lasts as long as it needs to, <u>until the working device has been constructed and tested.</u>

③ 英语中有的短语也可以起到时间状语连词的作用，这种从句在译成中文时要放在主句之前。常见的这类短语如下：

each time 每当　　every time 每当　　the moment…就

the minute…就　　by the time 到……时

<u>By the time protons come out of the cyclotron</u>, their speed has become nearly equal to that of light. 到质子离开回旋加速器时，其速度已接近于光速。

(2) 地点状语从句

<u>Where pipelines cross swamps, rivers, or lakes</u> they must be encased in concrete. 在管

道通过沼泽、河流或湖泊的地方，必须用混凝土把管道包起来。

Wherever you and I and the rest of the people on the earth go there is always gravity to keep us from falling off.

(3) 条件状语从句

① 条件状语可以放在句首、句中或句末，译成中文时一般放在句首。

If you were in a frictionless world, you would not be able to walk, no wheels would turn, and no vehicles could come to you. 如果处在一个没有摩擦的世界，你将无法走路，车轮将无法转动，也就不会有车来接你。

② 连词 as long as 及 so long as 译为"只要"。

A fan which is moving the air in a room helps us to feel cooler, **so long as** the air temperature is not warmer than that of the skin.

③ provided 及 providing (that)译为"如果，假如，只要"。如果译成"条件是"则从句须后置。

It is possible to establish an electric current in any material **provided** that sufficient voltage is available. 只要有足够的电压，就可以在任何材料中建立电流。

④ Unless 译为 "除非，如果不"。

Unless measures are taken at once, there is a good chance that all the oceans of the world be heavily polluted by the year 2000. 若不立即采取措施，全世界的海洋到 2000 年很可能受到严重的污染。

(4) 原因从句

Because aluminium is a good conductor of electricity it is often used for cables to carry telegraph and telephone messages. 因为铝是电的良导体，常常用作传送电报和电话信息的电缆。

As heat makes things move, it is a form of energy. 因为热能使物体运动，所以它是能量的一种形式。

Why can an airplane fly up **since it is much heavier than air**? 既然飞机比空气重得多，为什么飞机能飞上天空呢？

Now that man has left his footprints on the moon, where will he go next?

(5) 结果状语从句

The difficulties encountered have been **such that** little progress has yet been made. 所遇到的困难非常之大，以致进展甚微。

Most cells are **so** small **that** you need a powerful magnifying lens to see one.

(6) 目的状语从句

We have no protect the steel from contact with air when we heat or cool it **lest** oxidation should take place. 将钢加热或冷却时必须使之不与空气接触，以免发生氧化。

They use helium rather than hydrogen **for fear that** explosion should take place.

(7) 比较状语从句

Aluminum has a higher resistance to corrosion **than** many other metals. 铝的抗腐蚀能力**比**许多其他金属高。

Aluminum is **not so** expensive **as** copper or silver. 铝的价格**不如**铜或银**那样**贵。

The oxygen atom is nearly **16 times as** heavy **as** the hydrogen atom.

(8) 方式状语从句

Here on the earth we usually talk **as if** weight and mass were the same thing. 在地球上我们通常说起来**仿佛**重量和质量是一回事。

All active metals react slowly with oxygen in the air, **just as** iron does.

(9) 让步状语从句

Glass is a mixture of silicates, usually of sodium and calcium, **although** many variations are possible. 玻璃是硅酸盐的混合物，通常为硅酸钠和硅酸钙，**不过**也可能有许多其他品种。

Small **as** an atom is, it consists of smaller particles. 原子**尽管**很小，它**却**由更小的粒子组成。

Pure water, **no matter** what its source, can always be decomposed into oxygen and hydrogen.

五、科技英语中长句的特点及翻译

长句是科技英语的一大特点。造成句子长的原因有很多，但主要的原因是语言结构层次多而复杂：句子中并列成分多，各种短语、修饰语多，并列句或各种从句多。在翻译长句时，首先，不要因为句子太长而产生畏惧心理，因为，无论是多么复杂的句子，它都是由一些基本成分组成的。其次，要弄清英语原文的句法结构，找出整个句子的中心内容及其各层意思，然后分析几层意思之间的相互逻辑关系，再按照汉语的特点和表达方式，正确地译出原文，不必拘泥于原文的形式。

一般来说，造成长句的原因有三方面：①修饰语多；②并列成分多；③语言结构层次多。在分析长句时可以采用下面的方法。

① 找出全句的主语、谓语和宾语，从整体上把握句子的结构。

② 找出句子中所有的谓语结构、非谓语动词、介词短语和从句的引导词。

③ 分析从句和短语的功能，例如，是否为主语从句、宾语从句、表语从句等，若是状语，分析它是表示时间、原因、结果，还是表示条件等。

④ 分析词、短语和从句之间的相互关系，例如，定语从句所修饰的先行词是哪一个等。

⑤ 注意插入语等其他成分。

⑥ 注意分析句子中是否有固定词组或固定搭配。

例1：

Behaviorists suggest that the child who is raised in an environment where there are many stimuli which develop his or her capacity for appropriate responses will experience greater intellectual development.

分析：①该句的主语为 behaviorists，谓语为 suggest，宾语为一个从句，因此整个句子为 Behaviorist suggest that-clause 结构。

② 该句共有五个谓语结构，它们的谓语动词分别为 suggest, is raised, are, develop, experience 等，这五个谓语结构之间的关系为：Behaviorist suggest that-clause 结构为主句；who is raised in an environment 为定语从句，所修饰的先行词为 child；where there are many stimuli 为定语从句，所修饰的先行词为 environment；which develop his or her capacity for appropriate responses 为定语从句，所修饰的先行词为 stimuli；在 suggest 的宾语从句中，主语为 child，谓语为 experience，宾语为 greater intellectual development。

在做了如上的分析之后，我们就会对该句有了一个较为透彻地理解，然后根据上面所讲述的各种翻译方法，就可以把该句翻译成汉语。

译文：行为主义者认为，如果儿童的成长环境里有许多刺激因素，这些因素又有利于其适当反应能力的发展，那么，儿童的智力就会发展到较高的水平。

例2：

For a family of four, for example, it is more convenient as well as cheaper to sit comfortably at home, with almost unlimited entertainment available, than to go out in search of amusement elsewhere.

分析：①该句的骨干结构为 it is more… to do sth. than to do sth. else. 这是一个比较结构，而且是在两个不定式之间进行比较。

② 该句中共有三个谓语结构，它们之间的关系为：it is more convenient as well as cheaper to…为主体结构，但 it 是形式主语，真正的主语为第二个谓语结构 to sit comfortably at home，并与第三个谓语结构 to go out in search of amusement elsewhere 作比较。

③ 句首的 for a family of four 作状语，表示条件。另外，还有两个介词短语作插入语：for example，with almost unlimited entertainment available，其中第二个介词短语作伴随状语，修饰 to sit comfortably at home。

综合运用上述翻译方法，就可以把这个句子翻译为汉语。

译文：譬如，对于一个四口之家来说，舒舒服服地在家中看电视，就能看到几乎数不清的娱乐节目，这比到外面别的地方去消遣又便宜又方便。

此外，在进行翻译时，要特别注意英语和汉语之间的差异，将英语的长句分解，翻译成汉语的短句。在英语长句的翻译过程中，我们一般采取下列方法。

1. 顺序法

当英语长句的内容叙述层次与汉语基本一致时，可以按照英语原文的顺序翻译成汉语。下面我们列举几个实例来说明。

例 1：

But now it is realized that supplies of some of them are limited, and it is even possible to give a reasonable estimate of their "expectation of life", the time it will take to exhaust all known sources and reserves of these materials.

分析：该句的骨干结构为"It is realized that…"，it 为形式主语，that 引导主语从句以及并列的 it is even possible to…结构，其中，不定式作主语，the time …是 expectation of life 的同位语，进一步解释其含义，而 time 后面的句子是其定语从句。五个谓语结构表达了四个层次的意义：A. 可是现在人们意识到；B. 其中有些矿物质的储藏量是有限的；C. 人们甚至还可以比较合理地估计出这些矿物质"可望存在多少年"；D. 这些已知矿源和储量将消耗殆尽的时间。根据同位语从句的翻译方法，把第四层意义的表达作适当调整。

译文：可是现在人们意识到，其中有些矿物质的储藏量是有限的，人们甚至还可以比较合理地估计出这些矿物质"可望存在多少年"，也就是说，经过若干年后，这些矿物的全部已知矿源和储量将消耗殆尽。

例 2：

If such alloys possess other properties which make them suitable for die casting, they are obvious choices for the process, because their lower melting point will lead to longer die lives than would be obtained with alloys of higher melting points.

分析：本句由主句和四个从句组成。句法结构依次是：①if 引导的条件状语从句，其中包含 which 引导的定语从句修饰名词 properties；②主句；③because 引导的原因状语从句；④more…than 结构的比较状语从句。翻译时采用顺译法先翻译第①部分：如果这类合金具有使它们适于压铸的其他性能，其中定语从句译为"的"字结构的定语，即"使它们适于压铸的"。再翻译第②部分：它们显然可以被选来用于压铸。第③部分的原因状语从句：因为它们熔点较低，所以……可以延长压铸模寿命，注意翻译原因状语从句时采用了分译法，即将其主语部分译成一个分句。最后翻译第④部分：比起高熔点合金来，并将此插在原因状语从句中。

译文：如果这类合金具有使它们适于压铸的其他性能，它们显然可以被选来用于压铸，因为它们熔点较低，较高熔点合金可以延长压铸模寿命。

例 3：

This method of using "controls" can be applied to a variety of situations, and can be used to find the answer to questions as widely different as "Must moisture be present if iron is to rust?" and "Which variety of beans gives the greatest yield in one season?"

译文：这种使用参照物的方法可以应用于许多种情况，也能用来找到很不相同的问题的答案，如"铁生锈，是否必须有一定的湿度才行"和"哪种豆类一季的产量最高"。

2. 逆序法

英语有些长句的表达次序与汉语表达习惯不同，甚至完全相反，这时必须从后面开始翻译。下面我们列举几个实例来说明。

例1：

Aluminum remained unknown until the nineteenth century, because nowhere in nature is it found free, owing to its always being combined with other elements, most commonly with oxygen, for which it has a strong affinity.

分析：本句由一个主句、两个原因状语从句和一个定语从句，"Aluminum remained unknown until the nineteenth century"是主句，也是全句的中心内容，全句共有四个谓语结构，共有五层意思：A. 铝直到19世纪才被人发现；B. 由于在自然界找不到游离状态的铝；C. 由于它总是跟其他元素结合在一起；D. 最普遍的是跟氧结合；E. 铝跟氧有很强的亲和力。按照汉语的表达习惯通常因在前、果在后，这样我们可以逆着原文的顺序翻译。

译文：铝总是跟其他元素结合在一起，最普遍的是跟氧结合；因为铝跟氧有很强的亲和力，由于这个原因，在自然界找不到游离状态的铝。所以，铝直到19世纪才被人发现。

例2：

Scientists are learning a great deal about how the large plates in the earth's crust move, the stresses between plates, how earthquakes work, and the general probability that given place will have an earthquake, although they still cannot predict earthquakes.

分析：本句由一个主句、四个并列宾语（其中两个宾语从句、两个短语）、一个定语从句和一个让步状语从句组成。其具体句法结构依次是：①although 引导的让步状语从句；②宾语从句；③名词短语作宾语；④宾语从句；⑤定语从句修饰宾语 probability；⑥名词短语作宾语；⑦主句。整个句子主要结构的翻译采用倒译法，翻译的顺序从例句中所标明的序号可一目了然。先翻译①尽管科学家仍无法预测地震；再翻译四个并列宾语中的前三个②、③、④，地壳中的大板块如何运动，板块间的压力如何，地震如何发生；然后翻译带有定语从句的第四个宾语即⑤、⑥，某地区发生地震的一般概率为多少；最后翻译⑦即主句，他们了解得越来越多。

译文：尽管科学家仍无法预测地震，但对地壳中的大板块如何运动，板块间的压力如何，地震如何发生、某地区发生地震的一般概率为多少，他们了解得越来越多。

3. 分句法

有时英语长句中主语或主句与修饰词的关系并不十分密切，翻译时可以按照汉语多用短句的习惯，把长句的从句或短语化成句子，分开来叙述，为了使语意连贯，有时需要适当增加词语。下面我们列举几个实例来说明。

例1：

Television, it is often said, keeps one informed about current events, allows one to follow the latest developments in science and politics, and offers an endless series of programmes which are both instructive and entertaining.

分析：在此长句中，有一个插入语"it is often said"，三个并列的谓语结构，还有一个定语从句，这三个并列的谓语结构尽管在结构上属于同一个句子，但都有独立的意义，因此在翻译时，可以采用分句法，按照汉语的习惯把整个句子分解成几个独立的分句。

译文：人们常说，通过电视可以了解时事，掌握科学和政治的最新动态。从电视里还可以看到层出不穷、既有教育意义又有娱乐性的新节目。

例2：

The classical metallurgical processes of smelting the oxides with carbon in the presence of a fusible slag,①such as are used for the production of many of the commoner metals②, are not applicable to the range of rather rare elements about which this section is written③, if the metals are required in pure condition.④

分析：本句由一个主句、一个方式状语从句和一个条件状语从句组成。其中主句的主语与谓语被 such as 引导的方式状语从句隔开，主句后又有 about which 引导的定语从句修饰 rare elements，最后是 if 引导的条件状语从句。本句由于主句太长，适合采用分译法，即将主句分译成两个分句。具体的译法为：先翻译①，即主句的主语，把它译成一个分句，传统的冶金过程是用炭将易熔渣中的氧化物熔化；然后翻译②，许多普通金属都是这样生产的；再翻译③，即主句的谓语部分，将该部分译为另一个分句，但这种方法并不适用于生产本文所提到的这些稀有金属；最后翻译④，尤其是需要获得纯净金属时更是如此。

译文：传统的冶金过程是用炭将易熔渣中的氧化物熔化，许多普通金属都是这样生产的，但这种方法并不适用于生产本文所提到的这些稀有金属，尤其是需要获得纯净金属时更是如此。

4. 综合法

上面讲述了英语长句的逆序法、顺序法和分句法。事实上，在翻译一个英语长句时，并不只是单纯地使用一种翻译方法，而是要求我们把各种方法综合使用，这在我们上面所举的例子中也有所体现。尤其是在一些情况下，一些英语长句单纯采用上述任何一种方法都不方便，这就需要我们仔细分析，或按照时间顺序，或按照逻辑顺序，顺逆结合，主次分明地对全句进行综合分析，把英语原文翻译成通顺并符合原意的汉语句子。下面我们列举几个实例来说明。

例1：

People were afraid to leave their houses, for although the police had been ordered to stand by in case of emergency, they were just as confused and helpless as anybody else.

分析：该句共有三层含义，A. 人们不敢出门；B. 尽管警察已接到命令，要做好准备以应付紧急情况；C. 警察也和其他人一样不知所措和无能为力。在这三层含义中，B 表示让步，C 表示原因，而 A 表示结果。按照汉语习惯顺序，我们做如下翻译。

译文：尽管警察已接到命令，要做好准备以应付紧急情况，但人们仍不敢出门，因为警察也和其他人一样不知所措和无能为力。

例2:

As the science of gene expression grows①, we may be able to create genes ② that can turn themselves off ③ after they have gone through a certain number of cell divisions ④ or after the gene has produced a certain amount of the desired products ⑤.

分析：本句由一个主句、三个时间状语从句和一个定语从句组成，其中两个 after 引导的时间状语从句时修饰定语从句的谓语 turn off。本句总体采用分译法，即将主句② 和⑤分译为两个分句，具体的翻译方法为，先采用顺译法，翻译①即 as 引导的时间状语 从句，随着基因表现科学的发展；然后翻译②即主句，我们也许能够创造这样一些基因； 再采用倒译法，翻译③和④即两个 after 引导的时间状语从句，当它们经过了一定次数的 细胞分裂后，或者当它们产生了一定数量的合乎需要的产品之后；最后将定语从句⑤译 为一个独立的分句，这种基因能够自行衰亡。

译文：随着基因表达科学的发展，我们也许能够创造这样一些基因：当它们经过了 一定次数的细胞分裂后，或者当它们产生了一定数量的合乎需要的产品之后，这种基因 能够自行衰亡。

Excercises

Try to translate the following sentences.

① Water is usually considered as being a compound of two elements.

② Produced by electrons are the X-ray, which allow the doctor to see the inside of a patient's body.

③ It is said that all matters is made up of atoms. （主语从句）

④ Ancient people believed it to be true that the sun turned around the earth. （宾语从句）

⑤ Why don't we fall off the earth? The answer is that gravity keeps us from falling off. （表语）

⑥ We are all familiar with the fact that nothing in nature will either start or stop moving of itself. （同位语）

⑦ Fluids comprise both liquids and gases, the most common examples of which are water and air. （定语）

⑧ The body must dissipate heat as fast as it produces it. （状语）

⑨ Up to the present time, throughout the eighteenth and nineteenth centuries, this new tendency placed the home in the immediate suburbs, but concentrated manufacturing activity, business relations, government, and pleasure in the centers of the cities.

⑩ Although perhaps only 1 per cent of the life that has started somewhere will develop into highly complex and intelligent patterns, so vast is the number of planets that intelligent life is bound to be a natural part of the universe.

⑪ It therefore becomes more and more important that, if students are not to waste their opportunities, there will have to be much more detailed information about courses and more advice.

⑫ The number of the young people in the United States who can't read is incredible about one in four.

Unit 3   Writing a Scientific Paper in English

Introduction to scientific paper

A research paper is a form of written academic communication which can be employed to disseminate useful information and to share academic ideas with others. Most of the research papers are written for publication in journals or conference proceedings in one's field. Publication is one of the fastest ways for propagating ideas and for professional recognition and advancement. If you have a clear idea about the features and styles of academic articles, it will be easier for you to successfully get your paper published in the target journal or accepted by an international conference.

Features of academic papers

• The first of the features of an academic paper is the content. It is no more and no less than an objective and accurate account of a piece of research you did, either in the humanities, social sciences, natural sciences or applied sciences. It should not be designed to teach or to provide general background.

• The second feature is the style of writing for this purpose. Your paper should contain three ingredients: precise logical structure, clear and concise language, and the specific style demanded by the journal to which it will be submitted.

• The third, which is indeed a part of the second, is the system of documenting the sources used in writing the article. At every step in the process of writing, you must take into account the ideas, facts, and opinions you have gained from sources you have consulted.

• One of the most convenient features of academic articles is that they are divided into clearly delineated sections. This is helpful because you only have to concentrate on one section at a time. You can thus visualize more or less completely the whole paper while you are working on any part of it.

How to write a scientific paper

Although there is no fixed set of "writing rules" to be followed like a cookbook recipe or an experimental procedure, some guidelines can be helpful. Start by answering some questions:

• What is the function or purpose of this paper? Are you describing original and significant research results? Are you reviewing the literature? Are you providing an overview of the topic?

• How is your work different from that described in other reports on the same subject? (Unless you are writing a review, be sure that your paper will make an original contribution.)

• What is the best place for this paper to be published—in a journal or as part of a book? If a journal, which journal is most appropriate?

• Who is the audience? What will you need to tell them to help them understand your work?

• Answering these questions will clarify your goals and thus make it easier for you to write the paper with the proper amount of detail. It will also make it easier for editors to determine the paper's suitability for their publications. Writing is like so many other things: if you clarify your overall goal, the details fall into place.

• Once you know the function of your paper and have identified its audience, review your material for completeness or excess. Then, organize your material into the standard format: introduction, experimental details or theoretical basis, results, discussion, and conclusions. This format has become standard because it is suitable for most reports of original research, it is basically logical, and it is easy to use. The reason it accommodates most reports of original research is that it parallels the scientific method of deductive reasoning: define the problem, create a hypothesis, devise an experiment to test the hypothesis, conduct the experiment, and draw conclusions.

一、科技论文的组成（Elements of scientific papers）

① 前置部分：Title，Author/Affiliation，Abstract，Keywords；
② 主体部分：引言 Introduction，实验部分 Experimental，实验结果 Results，讨论 Discussion，结论 Conclusion；
③ 后置部分：致谢 Acknowledgement，参考文献 References，附录 Appendix。

二、题名（Title）

科技论文的题目又称文题、题目。它以最精练的文字，充分反映论文的基本内涵与特色，是论文内容的高度概括，也是反映论文特定内容的最恰当、最简明的词语逻辑组合。

1. 作用（Functions）

2. 题名的 ABC 三大原则

Accuracy 准确；Brevity 简练；Clarity 清晰。

3. 题目的书写

(1) 以名词短语为主要形式

• Inositol triphosphate and calcium signaling 三磷酸肌醇和钙信号表达

- Determination of Five Flavonoids in *Flos Inulae* 旋覆花中五种黄酮的测定
- Purification and identification of polysaccharide derived from *Chlorella pyrenoidosa* 蛋白核小球藻多糖的纯化和鉴定
- Adsorption of Calcium (Ⅱ) from aqueous solution by surface oxidized carbon nanotubes 用表面氧化碳纳米管吸收水溶液中的钙(Ⅱ)

(2) 题名通常不用陈述句及疑问句

① 陈述句容易使题名具有判断式的语义，同时，陈述句一般显得不是那么简练、清晰。

- Nitrendipine is Effective on Severe Hypertension 尼群地平对重症高血压有效
- Nitrendipine Being Effective on Severe Hypertension
- Nitrendipine Effective on Severe Hypertension

② 疑问句虽然是完整的句子，但属于探讨性的语气。

Is Nitrendipine Effective on Severe Hypertension?

(3) 慎重使用缩略语

在题名中，缩略语的使用必须严加限制——只有全称较长，缩写后的形式已经被公知公认，并且在读者群中非常流行时，才可使用。

- Determination of Puerarin in Jiangzhi Jianfei Granules by **HPLC** （HPLC—high performance liquid chromatographic）
- The effects of air pollutants on taking of SARS **GEE** （GEE—generalized estimating equation）

(4) 避免使用化学式、上下角标、特殊符号（数字符号、希腊字母等）、公式等

- On the Diophantine Equation $x^3 \pm 8 = Dy^2$
- Determination of α-Tocopherol and α-Tocopherol Nicotinate by HPLC
- Enantioselective Hydrogenation to Synthesize Ethyl (R)-2-Hydroxy-4-Phenylbutyrate

(5) 避免使用 Thoughts on..., Regarding…, Study…, 等

虽然不是不可用，但根据题名 ABC 原则，完全可以改写得更简练些。

(6) 题名书写格式

① 全部字母大写

DETERMINATION OF REACTOR KINETIC PARAMETERS IN A TWO-CORE REACTOR （双堆芯反应堆中反应堆动态参数的确定）

② 每个实词的首字母大写

Determination of Reactor Kinetic Parameters in a Two-core Reactor

③ 整个题名中，只有词的首字母大写，其余均小写

Determination of reactor kinetic parameters in a two-core reactor

(7) 举例

- 硅胶柱色谱法提纯大豆磷脂酰胆碱的研究

Purification of soybean phosphatidylcholine by silica gel column chromatography

- CPP 膜的等离子体表面改性

Plasma-modification on the surface of CPP film

- 大孔树脂对甘油的静态吸附及其热力学研究

Study on the static absorption behaviors and thermodynamic properties of glycerine on macroporous resins

- 桑根酮 C、桑根酮 D 在碱液中的稳定性研究

Study on the Stability of Sanggenon C and D in Base Solutions

三、作者及单位

1. 作者（Athor）

应尽量采用相对固定的英文姓名的表达形式，以减少在文献检索和论文引用中被他人误解的可能性。

表示方法（以"张三三"为例）：

① 国家标准（GB/T 16159—1996）：Zhang Sansan, Sansan Zhang；

② 中国学术期刊(光盘版)检索与评价数据规范：Zhang San-san, San-san Zhang；

③ 缩写：Zhang S., Zhang S.S., S.Zhang, S.S. Zhang。

2. 作者单位（Affiliation）

一般包括两个部分：一是单位名称，二是单位地址。例如：

- 台州学院医药化工学院，浙江临海 317000

School of Pharmaceutical and Chemical Engineering, Taizhou University, Linhai 317000, Zhejiang, China

- 河北医科大学药学院，（中国）石家庄 050017

School of Pharmacy, Hebei Medical University, Shijiazhuang 050017, China

四、关键词（Key words/keywords/Key Words）

As the name implies, keywords are the most important words and phases representative of the theme of the paper, and frequently used in a paper. Reader can find out the theme of the paper by looking at the keywords. The function of keywords is to facilitate the information retrieval and emphasize the gist of the paper.

- Limited Number: The number of keywords for a paper should be limited. Four to six keywords are the average. In general, there should be at least 2 and at most 8.

- Designated Choice: The keywords of a paper usually come from the title and/or the abstract, where the key terms of words and phrases are usually contained.

- Writing Requirements: Though keywords can be either above or below the abstract of a paper, they are yet, in most cases, placed below the abstract.

五、摘要

1. Linguistic Features of Abstract

An abstract is a miniature of the paper with a strictly limited number of words. Normally, 200 words should be a sensible maximum for a relatively long paper or report; 50-100 words may suffice for a short article. The length of an abstract greatly varies depending on the length of the paper. As a general rule, an abstract will be approximately 3%-5% of the length of the paper.

- A Statement of the Problem
- A Statement of the Approach to Solving the Problem
- The Principal Result
- Conclusions

2. Classification

(1) 陈述型摘要，又称指示型摘要 (indicative abstract)

只简单地介绍论文的论题，或者概括地表达研究目的，仅使读者对论文的主要内容有一个概括的了解。长篇论文有大量的数据，但是无法容纳在有限篇幅的摘要里。因此，一般不介绍方法、结果、结论的具体内容，不包含任何数据。通常陈述型摘要只有几个简单句，一般30～50个词。

This paper describes the simulation experiment of three kinds of plant steam pipe break accidents on Beijing nuclear power plant simulator, shows the curves of main operational parameters of the plant in three different conditions, analyses and discusses the experimental results. Also, this paper describes the general process on the accident treatment.

(2) 信息型摘要，又称资料型摘要

用来报道论文的研究目的、方法、结果与结论。实质上，它是整篇论文的高度浓缩，一般来讲，目前科技期刊中学术类文章的英文摘要大多属于信息型摘要，也是我们主要讨论的一种摘要。一般100～150个词。

A detailed study of the melting behavior of oriented isotactic polypropylene has been carried out using differential scanning calorimetry (差示扫描量热法, DSC). The orientation in isotactic polypropylene was produced by extruding (挤压) it in solid phase. At extrusion ratio (挤压比, ER) greater four, two melting peaks were observed. With increasing ER, the lower temperature peak was found to shift to higher temperature. The corresponding shift in the higher temperature peak was much less. It is shown that these peaks originate in the melting of crystalline species having different degrees of crystal disorder and stereo-block (立构嵌段) character.

3. How to write abstract for scientific paper

(1) 时态

从理论上讲，以一般现在时为主，也使用一般过去时和现在完成时。

一般现在时：通过科学实验取得的研究结果、结论，揭示自然界的客观规律。

一般过去时：在一定范围内所观察到的自然现象的规律性认识，这种认识也许有一定的局限性。

现在完成时：表明过程的延续性，虽某事件（或过程）发生在过去，但强调对现实所产生的影响。

EI 数据库建议： 用过去时态叙述作者工作；用现在时态叙述作者结论。

也有文献建议采用现在时：论文是通过科学实验揭示客观真理。所取得的结果，无论是过去、还是现在或将来都是如此。故常用现在时表达。过去时"表达一件过去发生过的事，而且现在已经完结了"。如文中指出发生的日期和时间是写文章之前，必须用过去时，例：This was first known in 1930. 许多论文，虽然是作者过去做的工作和得到的结论，然而这些工作和结论并不是达到"完结了"的阶段，而是还会有人，也可能是作者本人，继续研究下去，从而产生进一步的改进和完善。从这一角度出发，可以用现在完成时来描述已做过的工作，以表达这种延续性。例：Man has not yet discovered an effective cure for the common cold. （人类至今尚未发现一种有效治疗感冒的方法。）句中隐含着估计一段时间后也不会发现，但是将来能否会有，尚有待事实来说明。

以上是撰写摘要时常用的几种时态，有时很难区分它们在含义上的严格差异。这只是从语法功能的角度将其概念化，实际写作英文摘要时，这几种时态都可能存在，并非完全不可互易。例如：

① 一般现在时及其被动语态说明研究的目的、描述研究的内容、得出结果与结论等。

The aim of the determination is to study effects of radiates cotton cellulose by high energy electrons on properties of nitrocellulose (NC).

Under a certain nitrating condition, the nitrogen content and homogeneity of NC are retained basically with different doses of absorption.

② 一般过去时及其被动语态

An analysis of the cure kinetics of several different formulations composed of bifunctional epoxy resins and aromatic diamines was performed.

All kinetic parameters of the curing reaction were calculated and reported. Dynamic and isothermal DSC yielded different results. An explanation was offered in terms of different curing mechanisms which prevail under different curing conditions. A mechanism scheme was proposed to account for various possible reactions during cure.

Four kinds of liquid-liquid systems were examined.

Samples were extracted by 80% ethanol.

The flow rate was 1.0 ml·min^{-1} and detection was at 360nm.

③ 现在完成时及其被动语态

The partial molar enthalpies of mixing of $NaHSO_4$ and $KHSO_4$ have been measured at 528K by dropping samples of pure compounds into molten mixtures of $NaHSO_4$ and $KHSO_4$

in Calvet calorimeter. From these values the molar enthalpy of mixing has been deduced. The phase diagram of this system has been confirmed by conductometric and thermal analysis methods. By an optimization method the excess entropy of the liquid mixtures was also calculated.

(2) 摘要长度的控制

① 不能过于简单。

② 避免缩写字过多，不要使用图表或脚注、插图、表格、参考文献等有关符号。

③ 取消一些不必要的句子。

④ 可以对物理量单位以及一些通用词适当简化。

3km=three kilometers

UK=the United Kingdom

(3) 摘要的文体

① 英文摘要不同于文章正文，应尽量使用短句子，避免重复单调。

In this paper, an enhancement, which is named as ABC, is proposed to try to get a more powerful object-oriented method.→An enhancement method, ABC, is proposed. It is a more powerful object-oriented method.

② 可以用动词的情况应尽量避免使用动词的名词形式。

Measurement of thickness of plastic sheet was made→Thickness of plastic sheet was measured

③ 可直接用名词或名词短语作定语的情况下，要少用 of 句型。

accuracy of measurement→measurement accuracy

structure of crystal→crystal structure

④ 能用名词作定语的不用动名词作定语，能用形容词作定语的不用名词作定语。

measuring accuracy→measurement accuracy

experiment results→experimental results

⑤ 一个名词不宜用多个前置形容词来修饰，可改用复合词，兼用后置定语。

thermal oxidation apparent activation energy→apparent active energy of thermo-oxidation

the chlorine containing high melt index propylene based polymer→the chlorine-containing propylene-based polymer of high melt index…

⑥ 正确使用冠词，包括定冠词 the 和不定冠词 a 或 an。

⑦ 避免使用第一人称，以便于二次文献的编辑加工和利用。

⑧ 删繁就简。

at a temperature of 250℃ to 300℃→at 250-300℃

at a high temperature of 1 250℃→at 1 250℃

specially designed or formulated→nothing

has been found to increase→increased

from the experimental results, it can be concluded that→the results show was considered to be→was

it would seem that→seemingly

⑨ 单词拼写要统一，或用英式，或用美式，不要混用。（英→美）

travelling→traveling　　distil→distill　　colour→color　　storey→story

enclose→inclose　　　　flyer→flier　　　analyse→anlyze

programme→program

(4) 典型语句

1) 介绍文章作者的观点和研讨课题内容的语句

① 介绍文章内容与作者观点的常用语句

"deal with, describe, explain, illustrate, introduce, present, report, …"

The physical and chemical properties of the crystal are dealt with. 论及了此晶体的物理性质和化学性质。

The basic concepts and data collected are presented. 给出了基本概念与收集的数据。

② 文章研究课题的常用语句

"analyse, consider, develop, discuss, investigate, state, study, …"

Some applications of the two-phase flow in reactors are briefly discussed. 简要地讨论了反应堆里两相流的一些应用。

The relationship between mass and energy in nature is stated briefly. 简述了自然界质量与能量的关系。

③ 文章涉及范围的语句

"consist of, contain, cover, include, …"

So many elements such as Fe, Au, Ag, etc., are included in this project. 此项目中包括了像 Fe、Au、Ag 等许多元素。

Testing results of about 200 assemblies of fuel and control rods in the reactor core are contained. 包括了反应堆堆芯里约 200 个带燃料棒和控制棒的组件的测试结果。

④ 综述与概括对某一领域的研究课题的常用语句

"abstract, outline, review, summary"

The experimental technique for measurement of these properties is outlined only. 对测量这些特性的实验技术仅作概要叙述。

The theory based on materials science is summarized briefly. 简要地概述了基于材料科学的理论。

⑤ 文章重点的常用语句

"attention is concentrated on…, there is a focus on…, attention is paid to…, the emphasis is on…"

Attention is concentrated on extending the lead time to 24 hours. 重点是把交货时间延长到 24 小时。

The research is focused on collisions between neutrons and uranium nuclei. 集中研究中

子与铀原子核的碰撞。

⑥ 文章目的的常用语句

"aim, objective, purpose, seek"，经常要用动词不定式作表语或谓语动词的目的状语等。

The main objective of this study is to determine the nuclear properties of graphite as reactor moderators. 本研究的主要目标是要确定反应堆减速剂石墨的核特性。

The purpose of this research is to obtain the heavy concrete of much higher stability. 本项研究的目的是要获得具有很高稳定性的重混凝土。

2) 介绍文章成果的语句

① 成果的获取和开发等

"achieve, construct, derive, design, develop, establish, give, improve, obtain, produce, provide, realize, record, reduce, solve"

A dissolution diagram was constructed, which can be used in heat treatment. 绘制了用于热处理的溶解图。

Two methods of the boiler design are developed. 研发了锅炉设计的两种方法。

Good agreement with the corresponding results from the reactor dynamics is obtained. 得到了与相应反应堆动力学很一致的结果。

② 观察和指示等

"demonstrate, exhibit, fine, indicate, observe, point out, show"

All the salts studied, except NaCl, exhibit good catalytic activity. 所研究的盐类，除氯化钠外，都有很好的催化活性。

A great improvement in sensitivity and reproductivity is shown in the results by using this method. 这些结果表明，使用此方法能使灵敏度和再现性有很大的改善。

③ 运算和计算等

"calculate, determine, estimate, measure"

The earth-moon distance can be determined to an accuracy of ±1.5cm by using this technique. 用这种技术测得的地球与月亮之间的距离准确度可达±1.5cm。

The minimum fluidizing velocity at 1 000℃ was measured. 测量了1000℃时的最小流化速度。

④ 应用和用途等

"apply, use"

These facilities are used to develop computer hardware for several projects. 这些设施被用来为若干项目研制计算机硬件。

This method is used for measuring neutron flux distribution in the reactor core. 这种方法被用来测量反应堆堆芯的中子通量分布。

⑤ 评估与比较等

"agree with, assess, compare, evaluate"

These data are compared with those from the published experiments. 将这些数据与已发表的实验结果进行了比较。

The used assumptions are evaluated as compared with measurement results. 通过与测量结果比较，对所用的假设进行了评估。

⑥ 实验与试验等

"experiment, test"

Animals are experimented upon in the laboratory of the Beijing Zoo. 在北京动物园实验室里做动物实验。

The ascend rocket in the upper half of the lunar module was tested hundreds of times. 登月艇上半截的上升火箭，试验过几百次了。

⑦ 论证与依据等

"base on, be based on, take as reference"

Methods of analysis are based on Maxwell equations. 分析的方法是依据麦克斯韦方程式。

A simplified diagram of the reactor vessel and internals is taken as reference. 将反应堆容器与内部构件的简化图作为参考。

⑧ 推荐与建议等

"proposed, recommend, suggest, put forward"

A method is suggested to control and evaluate the corrosion of high temperature sections. 提出了一种方法来控制和评估高温区的腐蚀。

An alternative theory is put forward for the generation and migration of petroleum. 提出了石油生成和迁移的另一种理论。

⑨ 结论

"arrive at, conclude"

The main aim of engineering studies is to arrive at an optimum design. 工程研究的主要目的是要达到最优设计。

It is concluded that adding a nucleating agent and a thickening agent can inhibit supercooling and eliminate segregation. 结论是，加入成核剂和增稠剂能抑制过冷并消除分层。

Exercises

1. Put the following titles into English or Chinese

① 纯苦杏仁苷的制备（苦杏仁苷：amygdalin）

② HPLC-RI 法测定大豆磷脂酰胆碱含量（磷脂酰胆碱：phosphatidylcholine）

③ 聚丙烯酸钠高吸水性树脂的制备及性能研究

（丙烯酸钠：sodium acrylate；高吸水性：high water-absorbent；树脂：resin）

④ Removal of phosphate from aqueous solution by thermally treated natural palygorskite

⑤ Triblock Copolymer Syntheses of Mesoporous Silica with Periodic 50 to 300 Angstrom Pores

⑥ One-step synthesis of high capacity mesoporous Hg^{2+} adsorbents by non-ionic surfactant assembly

2. Put the following affiliation into English or Chinese

① 清华(Tsinghua)大学 信息科学技术学院 计算机科学与技术系，中国北京 100084

② 南京大学 污染控制与资源化国家重点实验室，江苏南京 210093

③ Institute of Pharmaceutical Engineering, College of Material and Chemical Engineering, Zhejiang University, Hangzhou 310027, Zhejiang, China

④ State Key Laboratory of Soil and Sustainable Agriculture, Institute of Soil Science, Chinese Academy of Sciences, 210008, Nanjing, China

⑤ Laboratory for Molecular Interfacial Interactions, Code 6930, Center for Bio/Molecular Science and Engineering, Naval Research Laboratory, Washington, DC 20375

⑥ Department of Chemistry and Biochemistry, University of California, Santa Barbara, CA 93106-9510, USA

3. Write your name in the three forms

Part II

Basic English Knowledges for Pharmaceutical and Chemical Engineerings

Unit 1　General Knowledge of Chemistry

Lesson 1　What is Chemistry?

① Roughly stated, physics is concerned with the general properties and energy and with events which results in what are termed physical changes. Physical changes are those in which materials are not so thoroughly altered as to be converted into other materials distinct from those present at the beginning.

② Chemistry, by contrast, is chiefly concerned with properties that distinguish materials from one another and with events which result in chemical changes. Chemical changes are those in which materials are transformed into completely different materials.

③ Such thoroughgoing transformations, in which all the properties of a material are altered, so that a completed different materials is obtained, are called chemical transformations, chemical changes or chemical reactions.

④ Chemistry as a science is a manner of thinking about transformations of materials which helps us to understand, predict and control them. It furnishes directing intelligence in the use of materials.

The scope of chemistry

① Chemistry is a science that tries to understand the properties of substances and the changes that substances undergo. It is concerned with substances that occur naturally—the minerals of the earth, the gases of the air, the water and salts of the seas, the chemicals found in living creatures and also with new substances created by humans. It is concerned with natural changes—the burning of a tree that has been struck by lightning, the chemical changes that are central to life, and also with new transformations invented and created by chemists.

② Chemistry is sometimes called the "central science" because it relates to so many areas of human endeavor and curiosity. Chemists who develop new materials to improve electronic devices such as solar cells, transistors, and optic cables work at the interfaces of chemistry with physics and engineering. Those who develop new pharmaceuticals for uses against cancer or AIDS work at the interfaces of chemistry with pharmacology and medicine.

③ Many chemists work in more traditional fields of chemistry. Biochemists are interested in chemical processes that occur in living organisms. Physical chemists work with fundamental principles of physics and chemistry in an attempt to answer the basic questions that apply to all of chemistry: Why do some substances react with one another while others do not? How fast will a particular chemical reaction occur? How much useful energy can be extracted from a chemical reaction?

④ Although chemistry is concerned a "mature" science, the landscape of chemistry is dotted with unanswered questions and challenges. Modern technology demands new materials with unusual properties, and chemists must devise new methods of producing these materials. Modern medicine requires drugs targeted to perform specific tasks in the human body, and chemists must design strategies to synthesize these drugs from simple starting materials. Society requires improved methods of pollution control, substitutes for scarce materials, nonhazardous means of disposing of toxic wastes, and more efficient ways to extract energy from fuels. Chemists are work in all these areas.

Fundamental principle of chemistry

① The first and most important principle is that chemical substances are made up of molecules in which atoms of various elements are linked in well-defined ways. The second principle is that there are somewhat more than 100 elements, which are listed in the periodic table of the elements. The third principle is that those elements, arranged according to increasing numbers of protons in their nuclei, have periodic properties.

② Another principle is that the ways in which atoms are linked strongly affect the properties of chemical substances. This is particularly evident when covalent links (bonds) are involved. Covalent bonds, in which two atoms are held together by a pair of electrons shared between them, are the bonds that hold the atoms of carbon, oxygen, and hydrogen together in cellulose, for instance. Most covalent bonds do not break easily, which is why intense heating is needed to turn cellulose into charcoal. The precise arrangement of the links determines chemical properties. By contrast, a salt such as sodium chloride has what are called ionic bonds. The sodium and the chlorine are not directly linked, just held together by the attraction of the positive sodium ion for the negative chloride ion. When sodium chloride is dissolved in water, the sodium ion and the chloride ion drift apart.

③ There is another much more subtle difference among chemical structures that has to do with the three-dimensional arrangement of atoms in space. Two chemicals can differ, even when all the same atomic linkages are present, if the spatial arrangements are different. Differences in spatial arrangement can have several aspects, but the most interesting has to do with handedness, or what chemists call chirality. When a carbon atom carries four different chemical groupings, there are two different ways they can be arranged. For example, in the amino acid alanine the central carbon atom carries four different groups: a hydrogen atom, a nitrogen atom, and two carbon atoms that differ in what is attached to them.

(selected from Chemistry: Today and Tomorrow)

Major branches of chemistry

① The body of knowledge about chemicals and chemical reactions is so vast that for convenience chemists have divided the study of chemistry into several major branches.

• Analytical chemistry: The study of what types of elements and compounds are present in a sample of matter – called qualitative analysis – and how much of each element and compound is present in a sample of matter – called quantitative analysis.

• Physical chemistry: The study of the scientific laws and theories that attempt to describe and explain the structure of matter, the chemical bonds that hold matter together, the changes that matter undergoes, and energy involved in these changes.

• Organic chemistry: The study of the properties and reactions of hydrocarbons, compounds derived from hydrocarbons that contain one or more other elements such as oxygen, nitrogen, sulfur, phosphorus, and chlorine.

• Inorganic chemistry: The study of all elements and the properties and reactions of the compounds not classified as organic compounds.

• Biochemistry: The study of the properties and reactions of compounds found in living organisms and those that are important to living organisms.

② Inorganic, organic, analytical, physical chemistry, and biochemistry are the main branches of chemistry, but it is possible to combine portions of them, or to elaborate on them in many ways. For example, Bioinorganic chemistry deals with the function of the metals that are present in living matter and that are essential to life. Polymer chemistry deals with the formation and behavior of such substances as rayon, nylon, and rubber. (Some people would include inorganic polymers such as glass and quartz.) Pharmaceutical chemistry is concerned with drugs: their manufacture, their composition, and their effects upon the body. Environmental chemistry, of course, deals with the composition of the atmosphere and the purity of water supplies—essentially, with the chemistry of our surroundings.

More about chemistry as the useful science

There is growing recognition of broad opportunities for the application of chemistry. Chemistry, interacting with other disciplines, provides the fundamental knowledge required to deal with many of society's needs. These include new materials for aerospace, automotive and electronic industries; basic data for the design of effective environmental controls; and the understanding of life processes important to agriculture and health care. A report by the National Academy of Science entitled "Opportunities in Chemistry" also called the "Pimentel Report"—helps us to crystallize our thoughts about the central importance of chemistry and chemical engineering and the wide bounds of their application.

Introduction to analytical chemistry

Analytical chemistry involves separating, identifying, and determining the relative amounts of the components making up a sample of matter. Qualitative analysis reveals the chemical identity of the analytes. Quantitative analysis tells us the relative amount of one or more of these analytes in numerical terms. Qualitative information is required before a quantitative analysis can be undertaken. A separation step is usually a necessary part of both qualitative and quantitative analyses.

1. Steps in a typical quantitative analysis

Selecting a method → sampling → preparing a laboratory sample → defining replicate samples → preparing solution of the sample → eliminating interferences → calibration and measurement → calculating results → evaluating results and estimating their reliability.

2. Classification of analysis methods

Analytical methods are often classified as being either classical or instrumental. This classification is largely historical with classical methods, sometimes called wet-chemical methods, preceding instrumental methods by a century or more.

3. Classical methods

In the early years of chemistry, most analyses were carried out by separating the components of interest (the analytes) in a sample by precipitation, extraction, or distillation. For qualitative analyses, the separated components were then treated with reagents that yielded products that could be recognized by their colors, their boiling or melting points, their solubilities in a series of solvents, their odors, their optical activities, or their refractive indexes. For quantitative analyses, the amount of analyte was determined by gravimetric or by titrimetric measurements. In gravimetric measurements, the mass of the analyte or some compound produced from the analyte was determined. In titrimetric procedures, the volume or mass of a standard reagent required to react with the analyte was measured.

4. Instrumental methods

Many of the phenomena that instrumental methods are based on have been known for a century or more. Their application by most chemists, however, was delayed by lack of reliable and simple instrumentation. In fact, the growth of modern instrumental methods of analysis has paralleled the development of the electronics and computer industries. Table 2 lists most of the characteristic properties that are currently used for instrumental analysis.

Table 2 Chemical and physical properties employed in instrumental methods

Characteristic properties	Instrumental methods
Emission of radiation	Emission spectroscopy (X-ray, UV, visible, electron, auger); fluorescence, phosphorescence, and luminescence (X-ray, UV, and visible)
Absorption of radiation	Spectrophotometry and photometry (X-ray, UV, visible, IR); photoacoustic spectroscopy; nuclear magnetic resonance and electron spin resonance spectroscopy
Scattering of radiation	Turbidimetry; nephelometry; Raman spectroscopy
Refraction of radiation	Refractometry; interferometry
Diffraction of radiation	X-ray and electron diffraction methods
Rotation of radiation	Polarimetry; optical rotary dispersion; circular dichroism
Electrical potential	Potentiometry; chronopotentiometry
Electrical change	Coulometry
Electrical current	Amperometry; polarography
Electrical resistance	Conductometry
Mass	Gravimetry (quartz crystal microbalance)
Mass-to-charge ratio	Mass spectrometry
Rate of reaction	Kinetic methods
Thermal characteristics	Thermal gravimetry and titrimetry; differential scanning colorimetry; differential thermal analyses; thermal conductometric methods
Radioactivity	Activation and isotope dilution methods

The first law of thermodynamics

Thermodynamics is a macroscopic science, and at its most fundamental level, is the study of two physical quantities, energy and entropy. Energy may be regarded as the capacity to do work, whilst entropy maybe regarded as a measure of the disorder of a system. Thermodynamics is particularly concerned with the interconversion of energy as heat and work. In the chemical context, the relationships between these properties may be regarded as the driving forces behind chemical reactions. Since energy is either released or taken in by all chemical and biochemical processes, thermodynamics enables the prediction of whether a reaction may occur or not without need to consider the nature of matter itself. Consideration of the energetics of a reaction is only one part of the story. Thermodynamics determines the potential for chemical change, not the rate of chemical change—that is the domain of chemical kinetics.

1. Internal energy

A fundamental parameter in thermodynamics is the internal energy denoted as U. This is the total amount of energy in a system, irrespective of how that energy is stored. Internal energy is the sum total of all kinetic and potential energy within the system.

2. Work

Work is the transfer of energy as orderly motion. In mechanical terms, work is due to energy being expended against an opposing force. The total work is equal to the product of the force and the distance moved against it. Work in chemical or biological system generally manifests itself in only a limited number of forms. Those most commonly encountered are pressure-volume (PV) work and electrical work.

3. Heat

Heat is the transfer of energy as disorderly motion as the result of a temperature difference between system and its surroundings. When energy is put into a system, there is usually a corresponding rise in the temperature of that system. Assuming that the energy is put in only as heat, then the rise in temperature of a system is proportional to the amount of heat which is input into it.

4. The first law

The first law of thermodynamics states that "The total energy of an isolated thermodynamic system is constant". The law is often referred to as the Conservation of Energy, and implies the popular interpretation of the first law, namely that "energy cannot be created or destroyed". In other words, energy may be lost from a system in only ways, either as work or as heat. As a result of this, it is possible to describe a change in the total internal energy as the sum of energy lost or gained as work and heat, since U cannot change in any other way. Thus, for a finite change: $\Delta U = Q + W$. Where Q is the heat supplied to the system, and W is the work done on the system.

Lesson 2 Chemical Bonding

Three difference types of strong of primary interatomic bonds are recognized: ionic, covalent, and metallic.

Ionic bonding: In the ionic bond, an electron donor (metallic) atom transfers one or more electrons to an electron acceptor (nonmetallic) atom. The two atoms then become a cation (e.g., metal) and an anion (e.g., nonmetal), which are strongly attached by the electostatic effect. This attraction of cations and anions constitutes the ionic bond.

In ionic solids composed of many ions, the ions are arranged so that each cation is surrounded by many anions as possible to reduce the strong mutual repulsion of cation. This packing further reduces the overall energy of the assembly and leads to a highly ordered arrangements called a crystal structure. The loosely bound electrons at the atoms are now tightly held in the locality of the ionic bond. Thus, the electron structure of the atom is changed by the ionic bond. in addition, the bound electrons are not available to serve as charge carriers and ionic solids are normally poor electrical conductors. Finally, the low overall energy state of these substances endows them with relatively low chemical reactivity. Sodium chloride (NaCl) is the classical ionic material (Figure 1). Sodium fluoride (NaF) and magnesium chloride ($MgCl_2$) are also example of ionic solids.

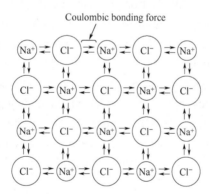

Figure 1 Schematic representation of ionic bonding in sodium chloride (NaCl)

Covalent bonding: Elements that fall along the boundary between metals and nonmetals, have atoms with four valence electrons and about equal tendencies to donate and accept electrons. For this reason, they do not form strong ionic bonds. Rather, stable electron structures are achieved by sharing valence electrons. For example, two carbon atoms can each contribute an electron to a shared pair. This shared pair of electrons constitutes the covalent bond. Covalent bonding is schematically illustrated in Figure 2 for a molecular of methane (CH_4).

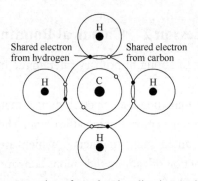

Figure 2　Schematic representation of covalent bonding in a molecule of methane (CH$_4$)

If a central carbon atom participates in four of these covalent bonds (two electrons per bond), it has achieved a stable outer shell of right valence electrons. More carbon atoms can be added to the grouping aggregate so that every atom has four nearest neighbors with which it shares one bond each. Thus, in a large grouping, every atom has a stable electon structure and four nearest neighbors. These neighbors often form a tetrahedron, and the tetrahedral in turn are assembled in an orderly repeating pattern (i.e., a crystal). This is the structure of both diamond and silicon. Diamond is the hardest of all materials, which shows that covalent bonds can be very strong. Once again, the bonding process results in a particular electronic structure (all electrons in pairs localized at the covalent bonds) and a particular atomic arrangement or crystal structure. As with ionic solids, localization of the valence electrons in the covalent bond renders these materials poor electrical conductors.

Metallic bonding: The third and least understood of the strong bond is the metallic bond. Metal atoms, being strong electron donors, do not bond by either ionic or covalent processes. Nevertheless, many metals are very strong and have high melting points, suggesting that very strong interatomic bonds are at work here, too. The model that accounts for this bonding envisions the atoms arranged in an orderly, repeating three-dimensional pattern, with the valence electrons migrating between the atoms like a gas (Figure 3).

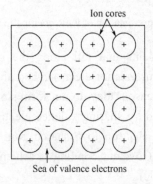

Figure 3　Schematic illustration of metallic bonding

It is helpful to imagine a metal crystal composed of positive ion cores, atoms without their valence electrons, about which the negative electrons circulate. On the average, all the electrical charges are neutralized throughout the crystal and bonding arises because the negative electrons act like a glue between the positive ion cores. This construct is called the free electron model of metallic bonding. Obviously, the bond strength increases as the ion cores and electron "gas" become more tightly packed (until the inner electron orbits of the ions begin to overlap). This leads to a condition of lowest energy when the ion cores are as close together as possible.

Once again, the bonding leads to a closely packed (atomic) crystal structure and a unique electronic configuration. The non-localized bonds within metal crystals permit plastic deformation (which strictly speaking does not occur in any nonmetals), and the electron gas accounts for the chemical reactivity and high electrical and thermal conductivity of metallic systems.

Weak bonding: In addition to the three strong bonds, there are several weak secondary bonds that significantly influence the properties of some solid materials, especially polymers. The most important of these are van der Waals bonding and hydrogen bonding, which have strengths 3%-10% that of the primary C-C covalent bond.

Atomic structure: The three-dimensional arrangement of atoms or ions in solid materials is one of the most important structural features that derive from the nature of the solid-state bond. In the majority of solid materials, this arrangement constitutes a crystal. A crystal is a solid whose atoms or ions are arranged in an orderly repeating patterns in three dimensions. These patterns allow the atoms to be closely packed (i.e., have the maximum possible number of near or contacting neighbors) so that the number of primary bonds is maximized and the energy of the aggregate is minimized.

Crystal structures are often represented by repeating electrons or subdivisions of the crystal called unit cells. Unit cells have all the geometric properties of the whole crystal. A model of the whole crystal can be generated by simply stacking up unit cells like blocks or hexagonal tiles.

Lesson 3 Working in Chemical Lab

Safety in the lab

Chemical experimentation, like driving a car or operating a household, creates hazards. The primary safety rule is to familiarize yourself with the hazards and then to do nothing that you (or your instructor or supervisor) consider to be dangerous. If you believe that an operation is hazardous, discuss it first and do not proceed until sensible precautions are in place.

Preservation of a habitable planet demands that we minimize waste production and responsibly dispose of waste that is generated. Recycling of chemicals is practiced in industry for economic as well as ethical reasons; it should be an important component of pollution control in your lab.

Before working, familiarize yourself with safety features of your laboratory. You should wear goggles or safety glasses with side shields at all times in the lab to protect your eyes from liquids and glass, which fly around when least expected. Contact lenses are not recommended in the lab because vapors can be trapped between the lens and your eyes. You can protect your skin from spills and flames by wearing a flame-resistant lab coat. Use rubber gloves when pouring concentrated acids. Never eat food in the lab.

Organic solvents, concentrated acids, and concentrated ammonia should be handled in a fume hood. Air flowing into the hood keeps fumes out of the lab. The hood also dilutes fumes with air before expelling them from the roof. Never generate large quantities of toxic fumes that are allowed to escape through the hood. Wear a respirator when handling fine powders, which could produce a cloud of dust that might be inhaled.

Clean up spills immediately to prevent accidental contact by the next person who comes along. Treat spills on your skin first by flooding with water. In anticipation of splashes on your body or in your eyes, know where to find and how to operate the emergency shower and eyewash. If the sink is closer than an eyewash, use the sink first for splashes in your eyes. Know how to operate the fire extinguisher and how to use an emergency blanket to extinguish burning clothing. A first aid kit should be available, and you should know how and where to seek emergency medical assistance.

Label all vessels to indicate what they contain. An unlabeled bottle left and forgotten in a refrigerator or cabinet presents an expensive disposal problem, because the contents must be analyzed before it can be legally discarded. National Fire Protection Association labels identify hazards associated with chemical reagents. A Material Safety Data Sheet provided with each chemical sold in the United States lists hazards and safety precautions for that chemical. It gives first aid procedures and instructions for handling spills.

Disposal of chemical waste

If carelessly discarded, many chemicals that we use are harmful to plants, animals, and people. For each experiment, your instructor should establish procedures for waste disposal. Options include ① pouring solutions down the drain and diluting with tap water, ② saving the waste for disposal in an approved landfill, ③ treating waste to decrease the hazard and then pouring it down the drain or saving it for a landfill, and ④ recycling. Chemically incompatible wastes should never be mixed with each other, and each waste container must be labeled to indicate the quantity and identity of its contents. Waste containers must indicate whether the contents are flammable, toxic, corrosive, or reactive, or have other dangerous properties.

A few examples illustrate different approaches to managing lab waste. Dichromate ($Cr_2O_7^{2-}$) is reduced to Cr^{3+} with sodium hydrogen sulfite ($NaHSO_3$), treated with hydroxide to make insoluble $Cr(OH)_3$, and evaporated to dryness for disposal in a landfill. Waste acid is mixed with waste base until nearly neutral (as determined with pH paper) and then poured down the drain. Waste iodate (IO_3^-) is reduced to I^- with $NaHSO_3$, neutralized with base, and poured down the drain. Waste silver or gold is treated to recover the metals. Toxic gases used in a fume hood are bubbled through a chemical trap or burned to prevent escape from the hood.

The lab notebook

The critical functions of your lab notebook are to state what you did and what you observed, and it should be understandable by a stranger. The greatest error, made even by experienced scientists, is writing incomplete or unintelligible notebooks. Using complete sentences is an excellent way to prevent incomplete descriptions.

Beginning students often find it useful to write a complete description of an experiment, with sections dealing with purpose, methods, results, and conclusions. Arranging a notebook to accept numerical data prior to coming to the lab is an excellent way to prepare for an experiment. It is good practice to write a balanced chemical equation for every reaction you use. This practice helps you understand what you are doing and may point out what you do not understand about what you are doing.

The measure of scientific "truth" is the ability of different people to reproduce an experiment. A good lab notebook will state everything that was done and what you observed and will allow you or anyone else to repeat the experiment.

Record in your notebook the names of computer files where programs and data are stored. Paste hard copies of important data into your notebook. The lifetime of a printed page is an order of magnitude (or more) greater than the lifetime of a computer disk.

Words and Expressions

experimentation   n. 实验，试验，实验法
hazard   n. 冒险，危险
preservation   n. 保存
habitable   adj. 可居住的
anticipation   n. 预期，预料
splash   n. 溅泼的斑点；v. 溅泼，溅湿
sink   n. 水槽，水池

extinguisher　n. 熄灭者，灭火器
goggle　n. 护目镜，（复数）风镜
spill　v. 溢出，溅出
eyewash　n. 洗眼水
emergency　n. 紧急情况，突然事件，非常时刻，紧急事件
assistance　n. 协助，援助，补助
precaution　n. 预防，防范
respirator　n. 口罩，防毒面具
inhale　v. 吸入，吸气
in anticipation (of)　预先，预料，期待
dilute with　用……稀释
unintelligible　adj. 难解的，无法了解的，莫名其妙的

Exercises

Put the following into English

矿物	分子	原子	质子
元素	周期表	有机化学	无机化学
分析化学	物理化学	生物化学	工业化学
化学工程	定性分析	定量分析	沉淀
萃取	蒸馏	沸点	熔点
折射率	光学活性	重量法	滴定法
热力学第一定律	熵	化学动力学	动能
势能	内能		

Unit 2　Nomenclature of Inorganic Compounds

1. The periodic table

The arrangement of elements in order of increasing atomic number, with elements having similar properties placed in vertical columns, is known as the periodic table.

The general structure of the modern periodic table will be familiar from previous chemistry courses. The elements are listed in order of atomic number, not atomic weight, for the atomic number tells us the number of electrons in the atom and is therefore a more fundamental quantity. The horizontal rows of the table are called periods and the vertical columns are called groups. We often use the group number to designate the general position of an element as in "gallium is in Group 13"; alternatively, the lightest element in the group is used to designate the group, as in "gallium is a member of the boron group". The members of the same groups as a given element are called the congeners of that element. Thus, sodium and potassium are congeners of lithium.

The periodic table is divided into four blocks. The members of the s- and p-blocks are collectively called the main-group elements, and the d-block elements (often with the exception of Group 12, zinc, cadmium, and mercury) are also referred to collectively as the transition elements. The f-block elements are divided into the lighter series (atomic numbers 57-71) called the lanthanides and the heavier series (atomic numbers 89~103) called the actinides. The representative elements are the members of the first three periods of the m group elements (from hydrogen to argon).

The numbering system of the groups is still in contention. The illustration shows both the traditional numbering of the main groups (with the roman numerals from Ⅰ to Ⅷ) and the current IUPAC recommendations, in which the groups of the s-, d-, and p-blocks are numbered from 1 through 18. The groups of the f-block are not numbered because there is little similarity between the lanthanides and the corresponding actinides in the period below.

2. Element names and symbols

Element—a fundamental building block of matter can not be broken down into simpler substances. It is most important that you begin to learn the names of the most common chemical elements along with the chemical symbols for them. Given the name of these elements, you should be able to write the symbols; given the symbols, you should be able to write the name. There is absolutely no way you can understand chemistry without learning this fundamental alphabet.

Chemical elements are symbolized by one- or two- letter abbreviations (缩写词) derived from their modern names, or in some cases from their old Latin names.

2.1　元素名称中词干的由来

Among 109 elements that have been named so far, their names come from a variety of sources. It will be helpful to remember their English names by knowing where the names of elements come from. In fact, most elements are named in honor of a scientist, a place or a country where an element was discovered. Since ancient Greece and ancient Rome were the important cradles of world civilization, the English names of many elements were derived from Greek or Latin. (详见：张文广，王祖浩. 无机物质的英文命名法[J]. 化学教育，2006, 8: 28-30)

2.2　常见元素的命名

Symbol	Chinese	Element	Symbol	Chinese	Element	Symbol	Chinese	Element
H	氢	hydrogen	Ne	氖	neon	Ag	银	silver/argentum
Br	溴	bromine	Mg	镁	magnesium	Sn	锡	tin/stannum
Li	锂	lithium	Al	铝	aluminum	Sb	锑	antimony/stibium
Zn	锌	zinc	Si	硅	silicon	Au	金	gold/aurum
B	硼	boron	P	磷	phosphorous	Hg	汞	mercury/hydrargyrum
I	碘	iodine	S	硫	sulfur	Pb	铅	lead/plumbum
Ba	钡	barium	Cl	氯	chlorine	Fe	铁	iron/ferrum
C	碳	carbon	Se	硒	selenium	K	钾	potassium/kalium
N	氮	nitrogen	Ra	镭	radium	Cu	铜	copper/cuprum
F	氟	fluorine	U	铀	uranium	Na	钠	sodium/natrium
Ca	钙	calcium	As	砷	arsenic	Mn	锰	manganese
Ti	钛	titanium	Cr	铬	chromium	Co	钴	cobalt

2.3　特殊元素的命名

Modern Name	Symbol	Derivation of Symbol	Modern Name	Symbol	Derivation of Symbol
sodium	Na	natrium	gold	Au	aurum
potassium	K	kalium	iron	Fe	ferrum
antimony	Sb	stibium	lead	Pb	plumbum
copper	Cu	cuprum	mercury	Hg	hydrargyrum
silver	Ag	argentum			

3. Inorganic Compounds

(1) Systematic chemical nomenclature

The systematic method of naming inorganic compounds considers the compound to be composed of two parts, one positive and one negative. The positive part, which is either a metal, hydrogen, or another positively charged group, is named and written first. The negative part, generally nonmetallic, follows. The names of the elements are modified with suffixes and prefixes to identify the different types or classes of compounds.

(2) Some common prefixes and their numerical equivalences

n	Prefix	n	Prefix	n	Prefix
1	mono	11	undeca	21	heneicosa
2	di	12	dodeca	22	doeicosa
3	tri	13	trideca	30	triaconsa
4	tetra	14	tetradeca	40	tetraconta
5	penta	15	pentadeca	50	pentaconta
6	haxa	16	hexadeca	60	hexaconta
7	hepta	17	heptadeca	70	heptaconta
8	octa	18	octadeca	80	octaconta
9	nona	19	nonadeca	90	nonaconta
10	deca	20	eicosa		

3.1 Binary compounds

Binary compounds contain only two different elements. Their names consist of two parts: the name of the more electropositive element followed by the name of the electronegative element, which is modified to end in *-ide*.

(1) Binary compounds in which the electropositive element has a fixed oxidation state

The major of these compounds contain a metal and a nonmetal. The chemical name is composed of the name of the metal, which is written first, followed by the name of the nonmetal, which has been modified to an identifying stem plus the suffix *-ide*. Stems of more common nonmetals are shown in the following table.

Table *Stems of more common negative-ion-forming elements*

Symbol	Element	Stem	Symbol	Element	Stem
B	boron	bor	I	iodine	iod
Br	bromine	brom	N	nitrogen	nitr
Cl	chloride	chlor	O	oxygen	ox
F	fluorine	fluor	P	phosphorus	phosph
S	sulfur	sulf			

Rule：某化某：阳离子元素名+盐成酸元素名词干+ide

Examples：

氯化钠	NaCl	sodium chloride	碘化钾	KI	potassium iodide
氮化镁	Mg$_3$N$_2$	magnesium nitride	溴化镁	MgBr$_2$	magnesium bromide
硫化钠	Na$_2$S	sodium sulfide	氯化铝	AlCl$_3$	aluminium chloride
氰化钠	NaCN	sodium cyanide			

(2) Binary compounds containing metals of variable oxidation numbers and nonmetals

Two systems are commonly used for compounds in this category. The official system, designated by the International Union of Pure and Applied Chemistry (IUPAC), is known as the Stock System. In the Stock System, when a compound contains a metal that can have more than one oxidation number, the oxidation number of the metal in the compound is designated by a Roman numeral in parentheses [e.g.,(Ⅱ)] written immediately after the name of the metal. The nonmetal is treated in the same manner as in the previous cases.

Examples: FeCl$_2$　　Iron (Ⅱ) chloride
　　　　　FeCl$_3$　　Iron (Ⅲ) chloride
　　　　　CuCl　　　Copper (Ⅰ) chloride
　　　　　CuCl$_2$　　Copper (Ⅱ) chloride

When a metal has only one possible oxidation state, there is no need to distinguish one oxidation state from another, so roman numerals are not needed.

In classical nomenclature, when the metallic ion has only two oxidation numbers, the name of the metal is modified with the suffixes **_-ous_** and **_-ic_** to distinguish between the two. The lower oxidation state is given the **_-ous_** ending, and the higher one is given the **_-ic_** ending.

Examples: FeCl$_2$　　Ferrous chloride

FeCl₃　Ferric chloride
CuCl　Cuprous chloride
CuCl₂　Cupric chloride

Names and oxidation numbers of some common metal ions that have more than one oxidation number.

Formula	Stock System name	Classical name	Formula	Stock System name	Classical name
Cu^+	copper(I)	cuprous	Sn^{4+}	tin(IV)	stannic
Cu^{2+}	copper(II)	cupric	As^{3+}	arsenic(III)	arsenous
Hg^+	mercury(I)	mercurous	As^{5+}	arsenic(V)	arsenic
Hg^{2+}	mercury(II)	mercuric	Sb^{3+}	antimony(III)	stibnous
Fe^{2+}	iron(II)	ferrous	Sb^{5+}	antimony(V)	stibnic
Fe^{3+}	iron(III)	ferric	Ti^{3+}	titanium(III)	titanous
Sn^{2+}	tin(II)	stannous	Ti^{4+}	titanium(IV)	titanic

Notice that the ous-ic naming system does not give the oxidation state of an element but merely indicates that at least two oxidation states exist. The Stock System avoids any possible uncertainty by clearly stating the oxidation number.

(3) Binary compounds containing two nonmetals

The chemical bond that exists between two nonmetals is predominantly covalent. In a covalent compound, positive and negative oxidation numbers are assigned to the elements according to their electronegativities. The most electropositive element is named first. In a compound between two nonmetals, the element that occurs earlier in the following sequence is written and named first: B, Si, C, P, N, H, S, I, Br, Cl, O, F.

To each element is attached a Latin or Greek prefix indicating the number of atoms of that element in the molecule. The second element still retains the modified binary ending. The prefix ***mono*** is generally omitted, except when needed to distinguish between two or more compounds, such as carbon monoxide, CO, and carbon dioxide, CO_2. Some common prefixes and their numerical equivalences as follows:

Mono=1　　Di=2　　Tri=3　　Tetra=4　　Penta=5
Hexa=6　　Hepta=7　　Octa=8　　Nona=9　　Deca=10

Generally the letter "*a*" is omitted from the prefix (from tetra on) when naming a nonmetal oxide. Examples of compounds illustrating this system are shown below.

CO	Carbon monoxide	N_2O	Dinitrogen oxide
CO_2	Carbon dioxide	N_2O_4	Dinitrogen tetroxide
PCl_3	Phosphorus trichloride	NO	Nitrogen oxide
PCl_5	Phosphorus pentachloride	N_2O_3	Dinitrogen trioxide

(4) Exceptions using -ide endings

Three notable exceptions using the ide ending are hydroxide (OH⁻), cyanide (CN⁻), and ammonium (NH_4^+) compounds. These polyatomic ions, when combined with another element, take the ending *ide*, even though more than two elements are present in the compound.

   NH_4I Ammonium iodide
   $Ca(OH)_2$ Calcium hydroxide
   KCN   Potassium cyanide

(5) Acids derived from binary compounds

Certain binary hydrogen compounds, when dissolved in water, form solutions that have acid property. Because of this property, these compounds are given acid names in addition to their regular ide names. For example, HCl is a gas and is called hydrogen chloride but its water solution is known as hydrochloric acid. Binary acids are composed of hydrogen and one other nonmetallic element. However, not all binary hydrogen compounds are acids. To express the formula of a binary acid, it is customary to write the symbol of hydrogen first, followed by the symbol of the second element (e.g., HCl, HBr, H_2S).

To name a binary acid, place the prefix hydro in front of and the suffix ic after the stem of the nonmetal, then add the word "acid".

Rule: 氢某酸: "hydro+成酸元素名词干+ic"acid

Examples:　HCl　Hydro chlor/ic acid (hydrochloric acid)
　　　　　　　　H_2S　Hydro sulfur/ic acid (hydrosulfuric acid)

氢氟酸	HF	hydrofluoric acid	F: fluorine
氢氯酸（盐酸）	HCl	hydrochloric acid	Cl: chlorine
氢溴酸	HBr	hydrobromic acid	Br: bromine
氢碘酸	HI	hydroiodic acid	I: iodine
氢氰酸	HCN	hydrocyanic acid	(CN): cyanogen
氢硫酸	H_2S	hydrosulfuric acid	S: sulfur
氢碲酸	H_2Te	hydrotelluric acid	Te: tellurium

3.2　Base

Classically, a base is a substance capable of liberating hydroxide ions, OH⁻, in water solution. Water solutions of bases are called alkaline solution or basic solutions. They have following properties: a bitter or caustic taste; a slippery, soapy feeling; the ability to change litmus from red to blue and the ability to interact with acids to form a salt and water.

Inorganic bases contain the hydroxyl group, OH⁻, in chemical combination with a metal ion. These compounds are called hydroxides. The OH⁻ group is named as a single ion and given the ending ide.

Rule: 氢氧化物：阳离子元素名+hydroxide

Examples: $Ca(OH)_2$ calcium hydroxide
  NaOH   sodium hydroxide $Al(OH)_3$ aluminum hydroxide
  NH_4OH ammonium hydroxide $Ba(OH)_2$ barium hydroxide

3.3 Acid

The word acid is derived from the Latin acids, meaning "sour" or "tart", and is also related to the Latin word acetum, meaning "vinegar". Vinegar has been know since antiquity as the product of the fermentation of wine and of apple cider. The sour constituent of vinegar is acetic acid ($HC_2H_3O_2$).

Some of the characteristic of properties commonly associated with acids are the following: Water solutions of acids are sour to the taste and are capable of changing the color of litmus, a vegetable dye, from blue to red. Water solutions of nearly all acids are able to react with: (i) metals such as zinc and magnesium to produce hydrogen gas; (ii) bases to produce water and a salt and (iii) carbonates to produce carbon dioxide. These properties are due to hydrogen ions, H^+, released by the acid in a water solution.

3.3.1 Acids derived from binary compounds [*See the (5) of 3.1*]

3.3.2 Ternary oxy-acids

(1) Ternary compounds

Ternary compounds contain three elements, and consist of an electropositive group, which is either a metal or hydrogen, and a polyatomic negative ion. The negative group usually contains two elements: oxygen and a metal or a nonmetal. In general, in naming ternary compounds the positive group is given first, followed by the name of the negative ion.

(2) Ternary oxy-acids

Inorganic ternary compounds containing hydrogen, oxygen, and one other element are called oxy-acids. The element other than hydrogen or oxygen in these acids is usually a nonmetal, but in some cases it can be a metal.

The ous-ic system is used in naming ternary acids. The suffixes ous and ic are used to indicate different oxidation states of the element other than hydrogen and oxygen. The ous ending indicates the lower oxidation state and the ic ending indicates the higher oxidation state. If an element has only one usual oxidation state, the ic ending is used.

In cases where there are more than two oxy-acids in a series, the ous-ic names are further modified with the prefixes per- and hypo-. Per is placed before the stem of the element other than hydrogen and oxygen when the element has a higher oxidation number than in the ic acid. Hypo is used as a prefix before the stem when the element has a lower oxidation number than in the ous-acid.

① 原酸或正酸。
Rule: "成酸元素名词干+ic"acid
Examples:

硫酸	H_2SO_4	sulfuric acid	S: sulfur=sulphur
硝酸	HNO_3	nitric acid	N: nitrogen
磷酸	H_3PO_4	phosphoric acid	P: phosphorus
氯酸	$HClO_3$	chloric acid	Cl: chlorine

② 亚酸。
Rule: "成酸元素名词干+ous"acid
Examples:

亚硫酸	H_2SO_3	sulfurous acid
亚硝酸	HNO_2	nitrous acid
亚磷酸	H_3PO_3	phosphorous acid

③ 高酸。
Rule: "per+成酸元素名词干+ic"acid
Examples:

高氯酸	$HClO_4$	perchloric acid	Cl: chlorine
高碘酸	HIO_4	periodic acid	I: iodine
高锰酸	$HMnO_4$	permanganic acid	Mn: manganese

④ 次酸。
Rule: "hypo+成酸元素名词干+ous"acid
Examples:

次磷酸	H_3PO_2	hypophosphorous acid
次氯酸	$HClO$	hypochlorous acid

⑤ 其他。
- 焦酸: "pyro+成酸元素名词干+ic"acid
  焦硫酸　　$H_2S_2O_7$　　pyrosulfuric acid
  焦磷酸　　$H_4P_2O_7$　　pyrophosphoric acid

- 偏酸："meta+成酸元素名词干+ic"acid
  偏磷酸　　　　　HPO$_3$　　　　　metaphosphoric acid
- 过氧酸："peroxy+成酸元素名词干+ic"acid
  过硫酸　　　　　H$_2$SO$_5$　　　　　peroxysulfuric acid
  过二硫酸　　　　H$_2$S$_2$O$_8$　　　　peroxydisulfuric acid
- 硫代酸："thio+成酸元素名词干+ic"acid
  硫代硫酸　　　　H$_2$S$_2$O$_3$　　　　thiosulfuric acid
  硫代氰酸　　　　HSCN　　　　　thiocyanic acid

4. Salt

When the hydrogen of an acid is replaced by a metal ion or an ammonium ion, the compound formed is classified as a salt. Therefore, we can have a series of metal chlorides, bromides, sulfides, sulfates, nitrates, phosphates, borates, and so on.

① Binary compounds in which the electropositive element has a fixed oxidation state [*See the (1) of 3.1*].

② Binary compounds containing metals of variable oxidation numbers and nonmetals[*See the (2) of 3.1*].

③ Ternary compound with only one positive ion.

To name the polyatomic negative ion, add the endings ite or ate to the stem of the element other than oxygen. Note that oxygen is not specifically included in the name, but is understood to be present when the endings ite and ate are used. The suffixes ite and ate represent different oxidation state. The ite ending refers to the lower oxidation state, while the ate ending represents the higher oxidation state. When an element has only one oxidation state, the ate ending is used.

硝酸钠	NaNO$_3$	sodium nitrate	高锰酸钾	KMnO$_4$	potassium permanganate
亚硝酸钠	NaNO$_2$	sodium nitrite	亚磷酸钾	K$_3$PO$_3$	potassium phosphite
硫酸铁	Fe$_2$(SO$_4$)$_3$		Iron(Ⅲ) sulfate or ferric sulfate		
硫酸亚铁	FeSO$_4$	Iron(Ⅱ) sulfate or ferrous sulfate			
硝酸铜	Cu(NO$_3$)$_2$	copper(Ⅱ) nitrate or cupric nitrate			
硝酸亚铜	CuNO$_3$	cupper(Ⅰ) nitrate or cuprous nitrate			

④ Ternary compound with more than one positive ion.

Salts may be formed by acids that contain two or more acid hydrogen atoms by replacing only one of the hydrogen atoms with a metal or by replacing both hydrogen atoms with different metals. Each positive group is named first and then the appropriate salt ending is added. The prefix bi- is commonly used to indicate a compound in which one of two acid hydrogen atoms has been replaced by a metal.

NaKSO$_4$	sodium potassium sulfate
MgNH$_4$SO$_4$	magnesium ammonium phosphate
NaHCO$_3$	sodium bicarbonate or sodium hydrogen carbonate
NaHS	sodium bisulfide or sodium hydrogen sulfide
KHSO$_4$	potassium bisulfate or potassium hydrogen sulfate
Al(HCO$_3$)$_3$	aluminum bicarbonate or aluminum hydrogen carbonate

Exercises

Write the English name of following compounds

NO	HCN	Mg(OH)$_2$	CuSO$_3$
NO$_2$	HI	NH$_4$OH	CuSO$_4$
N$_2$O	HClO	KOH	Cu$_2$SO$_4$
N$_2$O$_3$	HClO$_2$	CaC$_2$	CaCO$_3$
N$_2$O$_4$	HClO$_3$	AlCl$_3$	Ca(HCO$_3$)$_2$
N$_2$O$_5$	HClO$_4$	HgO	KAl(SO$_4$)$_2$

Unit 3 Nomenclature of Organic Compounds

1. Saturated hydrocarbons and their radicals

Hydrocarbons are composed entirely of carbon and hydrogen atoms bonded to each other by covalent bonds. They include the alkanes, alkenes, alkynes, and aromatic hydrocarbons. It is necessary to learn the names of the first 10 members of the alkane series as these names are used, with slight modifications, for corresponding compounds belonging to other classes.

Alkanes are acyclic hydrocarbons of general formula C_nH_{2n+2}.

The carbon atoms are arranged in chains that are either branched or unbranched.

$$CH_3-CH_2-CH_2-CH_3 \qquad \begin{array}{c}H_3C\\H_3C\end{array}\!\!\!>\!CH-CH_3$$

an unbranched alkane   a branched alkane

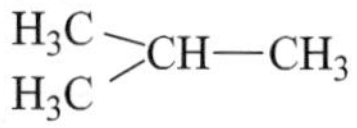

(1) Saturated unbranched acyclic hydrocarbons

The first four saturated unbranched acyclic hydrocarbons are called methane, ethane, propane and butane. Names of the higher members of this series consist of a numerical term, followed by *"-ane"* with elision of terminal *"a"* from the numerical term.

Rule: alkane= "Numerical term" + ane

The suffix "…ane" indicates that the compound is an alkane.

The prefix indicates the number of carbons in the compound.

Examples:

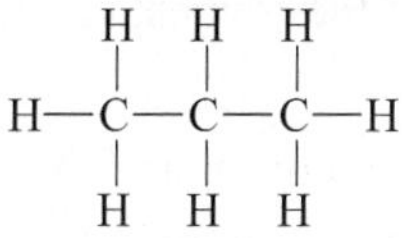

C_3H_8 丙烷 propane

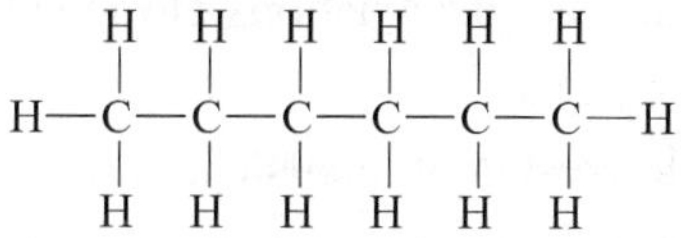

C_6H_{14} 己烷 hexane

Other alkanes:

n	nomenclature	n	nomenclature
1	methane	11	undecane
2	ethane	12	dodecane
3	propane	13	tridecane
4	butane	14	tetradecane
5	pentane	15	pentadecane
6	hexane	16	hexadecane
7	heptane	17	heptadecane
8	octane	18	octadecane
9	nonane	19	nonadecane
10	decane		

***Training*:**

C$_7$H$_{16}$ CH$_3$-[CH$_2$]$_{58}$-CH$_3$ CH$_3$-[CH$_2$]$_{130}$-CH$_3$

_____   _____   dotriacontahectane 一百三十二烷

(2) Alkyl group: univalent radicals

• Univalent radicals derived from saturated unbranched acylic hydrocarbons by removal of hydrogen from a terminal carbon atom are called normal or unbraced chain, alkyl.

• Alkyl radicals have the general formula C_nH_{2n+1}.

• The name of the radical is formed from the name of the corresponding alkane by simply dropping "-ane" and substituting a "-yl" ending.

Rule: "-ane" → "-yl"

Examples: 丁烷 CH$_3$-CH$_2$-CH$_2$-CH$_3$ butane → 丁基 CH$_3$-CH$_2$-CH$_2$-CH$_2$- butyl

名称	命名	构造式
甲基	methyl	—CH$_3$
乙基	ethyl	—CH$_2$CH$_3$
丙基	propyl	—CH$_2$CH$_2$CH$_3$
丁基	butyl	—CH$_2$(CH$_2$)$_2$CH$_3$
戊基	pentyl	—CH$_2$(CH$_2$)$_3$CH$_3$
己基	hexyl	—CH$_2$(CH$_2$)$_4$CH$_3$

(3) Saturated branched-chain compound

Rule1: prefixing the designations of the side chains to the name of the longest chain present in the formula.

i. Select the longest continuous chain of carbon atoms as the parent compound, and consider all alkyl attached to it as branch chains that have replaced hydrogen atoms of the parent hydrocarbon.

ii. Number the carbon atoms in the parent carbon chain from the end that gives the lowest numbers to the branch chains.

iii. Name each branch chain alkyl group and designate its position on the parent carbon chain by a number.

iv. When the same alkyl group branch chain occurs more than once, indicate it by a prefix, di-, tri-, tetra-, and so forth, written in front of the alkyl group name. The numbers indicating the positions of these alkyl groups are placed in front of the name.

v. If two or more side chains of different nature are present, they are cited in alphabetical order, regardless position number.

Example:

I	CH₃—CH₂—C(CH₃)(H)—CH₂—CH₂—C(CH₃)(CH₂—CH₂—CH₃)—CH₂—CH₃
II	CH₃—CH₂—C³(CH₃)(H)—CH₂—CH₂—C⁶(CH₃)(CH₂—CH₂—CH₃)—CH₂—CH₃ nonane (numbered 1,2,3,4,5,6,7,8,9)
III	Branches at C3 and C6——3-methyl, 6-methyl, 6-ethyl
IV	3-methyl, 6-methyl →3,6-dimethyl
V	6-ethyl -3,6-dimethylnonane

Training1:

CH₃—CH₂—CH(CH₃)—CH₃   CH₃—C(CH₃)(CH₃)—CH₂—C(CH₃)(H)—CH₃   

_____   _____   5,5-bis(1,2-dimethylpropyl)nonane

Rule2: The following names are retained for unsubstituted hydrocarbons only.

CH₃—CH₂—CH₂—CH₂—CH₃   CH₃—CH₂—CH(CH₃)—CH₃   CH₃—C(CH₃)(CH₃)—CH₃

正戊烷 *n*-pentane   异戊烷 *iso*-pentane   新戊烷 *neo*-pentane

CH₃—CH₂—CH(—)—CH₃   CH₃—C(CH₃)(CH₃)—   CH₃—C(CH₃)(CH₃)—CH₂—

仲丁基 *sec*-butyl   叔丁基 *tert*-butyl   新戊基 *neo*-pentyl

(4) cyclic hydrocarbons 环烷烃
- Alkanes can also form cyclic structures, general formula for cycloalkanes: C_nH_{2n}
- Name them, by a prefix "cyclo-"

cyclopropane   cyclohexane   _____

2. Unsaturated compounds and univalent radicals

(1) Alkenes

Unsaturated unbranched acyclic hydrocarbons having one double bond are named by replacing the ending "-ane" of the name of the corresponding saturated hydrocarbon with the ending "-ene". The chain is so numbered as to give the lowest possible numbers to the double bonds. If there are two or more double bonds, the ending will be "-adiene", "-atriene", etc. The generic names of these hydrocarbons (branched or unbranched) are "alkene", "alkadiene", "alkatriene", etc.

Example:

CH_3—CH_2—CH—CH—CH_3　　　pentane
CH_3—CH_2—CH=CH—CH_3　　　2-pentene (not 3-pentene) or pent-2-ene
CH_3=CH—CH=CH—CH_3　　　1,3-pentadiene or penta-1,3-diene

(2) Alkynes

Unsaturated unbranched acyclic hydrocarbons having one triple bond are named by replacing the ending "-ane" of the name of the corresponding saturated hydrocarbon with the ending "-yne". The chain is so numbered as to given the lowest possible numbers to the triple bonds. If there are two or more triple bonds, the ending will be "-adiyne", "-atriyne", etc. The generic names of these hydrocarbons (branched or unbranched) are "alkyne", "alkadiyne", "alkatriyne", etc. Only the lowest locant for a triple bond is cited in the name of a compound.

Example:

CH_3—CH_2—CH—CH—CH_3　　　pentane
CH_3—CH_2—C≡C—CH_3　　　2-pentyne (not 3-pentyne) or pent-2-yne
HC≡C—C≡C—CH_3　　　1,3-pentadiyne or penta-1,3-diyne

(3) Both have double bond and triple bond

The presence of both double and triple bonds is similarly denoted by ending such as "-enyne", "-adienyne", "-enediyne", etc. Numbers as low as possible are given to double bond and triple bonds as a set, even though this may at times given "-yne" a lower number than "-ene". If a choice remains, preference for two locants is given to the double bonds. The lower locant for a multiple bond is cited.

Example:

HC≡C—CH_2—CH=CH_2　　　pent-1-en-4-yne or 1-penten-4-yne
HC≡C—CH=CH—CH_3　　　pent-3-en-1-yne (not pent-2-ene-4-yne) or 3-penten-1-yne
HC≡C—CH=CH—CH=CH_2　　　1,3-hexadiene-5-yne or hexa-1,3-dien-5-yne

(4) Univalent radicals derived from unsaturated acyclic hydro carbons

The names of univalent radicals derived from unsaturated acyclic hydrocarbons have the ending "-enyl", "-ynyl", "-dienyl", etc. The position of the double and triple bonds being indicated where necessary, the carbon atom with the free valence is numbered as 1.

Example:

HC≡C—	ethynyl	$H_2C=CH-CH_2-CH_2-$	3-butenyl
HC≡C—CH$_2$—	2-propynyl	$H_2C=CH-CH=CH-$	1,3-butadienyl
HC≡C—CH=CH—CH$_2$—	2-penen-4-ynyl		

(5) Unsaturated branched acyclic hydrocarbons

Unsaturated branched acyclic hydrocarbons are named as derivatives of the unbranched hydrocarbons which contain the maximum number of double and triple bonds. If there are two or more chains competing for selection as the chain with the maximum number of unsaturated bonds, then the choice goes to: (i) that one with the greatest number of carbon atoms; (ii) the number of carbon atoms being equal, that one containing the maximum number of double bonds. In other respects, the same principles apply as for naming saturated branched acyclic hydrocarbons. The chain is so numbered as to give the lowest possible numbers to double and triple bonds.

Example:

$$H_2C=C-CH_2-CH_2-CH_2-CH_3$$
$$|$$
$$C_2H_5$$

2-ethyl-hexene or 2-ethyl-1-hexene

$$H_2C=CH-C=CH_2$$
$$|$$
$$C_6H_{13}$$

2-hexylbuta-1,3-diene or 2-hexyl-1,3-butadiene

(6) Unsaturated cyclic hydrocarbons

cyclopentene 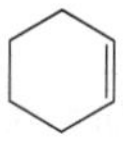   cyclohexene 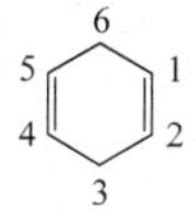   1,4-cyclohexadiene

(7) Some non-systematic names are retained

$H_2C=CH_2$	$H_2C=C=CH_2$	HC≡CH
thene→ethylene	propadiene→allene	ethyne→acetylene
$H_2C=CH-$	$H_2C=CH-CH_2-$	$H_2C=\overset{\mid}{C}-CH_3$
ethenyl→vinyl	2-propenyl→allyl	1-methylvinyl→isopropenyl

3. Aromatic compounds and their radicals

• The following names for monocyclic substituted aromatic hydrocarbons are retained.

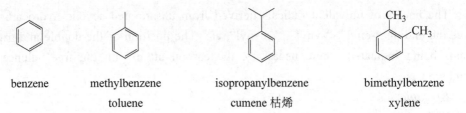

benzene	methylbenzene	isopropanylbenzene	bimethylbenzene
	toluene	cumene 枯烯	xylene

• The position of substituents is indicated by numbers except that *o*-(*ortho*), *m*-(*meta*) and *p*-(*para*) may be used in place of 1,2-, 1,3-, and 1,4-, respectively, when only two substituents are present.

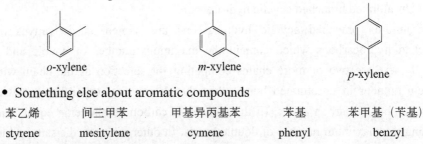

o-xylene   *m*-xylene   *p*-xylene

• Something else about aromatic compounds

苯乙烯	间三甲苯	甲基异丙基苯	苯基	苯甲基（苄基）
styrene	mesitylene	cymene	phenyl	benzyl

4. Other organic compounds with functional groups

(1) 醇 alcohol

① Alcohols may be oxidized to aldehydes, ketones, and carboxylic acids. Primary alcohols yield aldehydes and carboxylic acids; secondary alcohols are oxidized to ketones; tertiary alcohols resist oxidation.

② To name and alcohol by the IUPAC System:

• Select the longest continuous chain of carbon atoms containing the hydroxyl group.

• Number the carbon atoms in this chain so that the one bearing the –OH group has the lowest possible number.

• From the alcohol name by dropping the final—e from the corresponding alkane name and adding—ol. Locate the –OH group by putting the number (hyphenated) of the carbon atom to which it is attached immediately before the alcohol name.

• Name each alkyl side chain (or other group) and designate its position by number.

•脂肪族醇：将母体烃结尾的"e"改成"ol"

丁醇	3-甲基丁醇	2-丙醇	2-丙炔-1-醇
butanol	3-methylbutanol	2-propanol	2-propynyl-1-ol

•一些习惯命名

HO—CH$_2$—CH$_2$—CH$_2$—CH$_3$ CH$_3$—CH$_2$—CH(OH)—CH$_3$ HO—C(CH$_3$)$_2$—CH$_3$

正丁醇 *n*-butyl alcohol   仲丁醇 *sec*-butyl alcohol   叔丁醇 *ter*-butyl alcohol

- 多元醇:"e"保留,"ol"前加羟基数

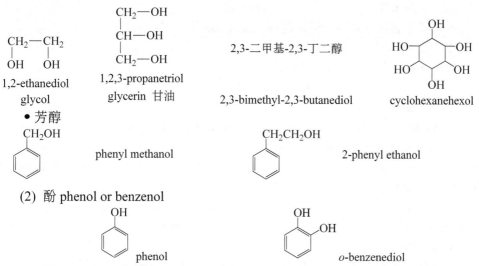

1,2-ethanediol glycol

1,2,3-propanetriol glycerin 甘油

2,3-二甲基-2,3-丁二醇 2,3-bimethyl-2,3-butanediol

cyclohexanehexol

- 芳醇

phenyl methanol

2-phenyl ethanol

(2) 酚 phenol or benzenol

phenol

o-benzenediol

(3) 醚 ethers

① Alcohols (ROH) and ethers (ROR) are isomeric, having the same molecular formula, but different structural formulas.

② The names of ethers consist of the names of the two radical groups attached to the oxygen followed by the word "ether".

- 系统命名法:将较长的烃基作为母体,余下的烷氧基作为取代基: -ane 换为-oxy

CH$_3$O—	甲氧基	methoxy
CH$_3$OCH$_2$CH$_3$	甲氧基乙烷	methoxyethane
CH$_2$CH$_2$OCH$_2$CH$_2$Br	1-溴-2-乙氧基乙烷	1-bromo-2-ethoxyethane

- 习惯命名法:按氧原子所连接的两个烃基的名称命名

CH$_3$—O—CH$_2$CH$_3$	甲基乙基醚	ethyl methyl ether
C$_2$H$_5$—O—C$_2$H$_5$	乙醚	ethyl ether
CH$_3$—O—C(CH$_3$)$_2$	甲基异丙醚	methyl isopropyl ether

(4) 醛 aldehyde

① Two classes of organic compounds that are similar in structure are aldehydes and ketones.

② Aldehydes are named by dropping the e in the parent hydrocarbon name and adding the letter al. The first member of the series is methanal. Its common name is formaldehyde.

- 系统命名法:将烃词尾"e"换为"al"

HCHO	methanal
CH$_3$CH$_2$CH$_2$CHO	butanal

$CH_2=CHCH_2CH_2CHO$　　　　pentenal
$CH\equiv C-CHO$　　　　　　propynal

- 习惯命名法（适用于低级醛）：-aldehyde

HCHO　　　　　formaldehyde
CH_3CHO　　　　acetaldehyde
$C_6H_5\text{-}CHO$　　　benzaldehyde

(5) 酮 ketone

① In the IUPAC system, ketones are named by dropping the "e" in the parent hydrocarbon name and adding the letters "one". The first member of the homologous series has three carbon atoms and is called propanone. Its common names are acetone and dimethyl ketone.

② When the carbon chain contains five or more carbon atoms, the longest chain containing the ketone group is numbered so that the ketone group will have the smallest possible number.

- 系统命名法：将烃词尾的"e"换为"one"；二酮、三酮"e"要保留

CH_3COCH_3　　　丙酮　　propanone
$CH_3CH_2COCH_3$　丁酮　　butanone

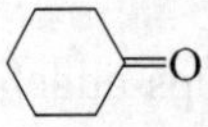

　　　　　　　　　环己酮
　　　　　　　　　cyclohexanone

- 习惯命名法：-ketone

CH_3COCH_3　　二甲基酮 dimethylketone　　　丙酮 acetone
$CH_3COC_2H_5$　甲基乙基甲酮 methyl ethyl ketone　乙基甲基酮 ethyl methyl ketone

(6) 酸 acid

① Organic acids, known as carboxylic acids, are characterized by the functional group called a carboxyl group.

② Carboxylic acids are named by dropping the e of the parent hydrocarbon name and adding the letters oic followed by the word acid. Thus the one carbon acid, HCOOH, is called methanoic acid.

③ Many carbonxylic acids are found in nature and their common names often reflect the natural source. Animal and vegetable fats and oils are important sources of organic acids that range up to eighteen carbon atoms [for example, stearic acid, $CH_3(CH_2)_{16}COOH$].

- 系统命名法：结尾去"e"加"-oic acid"

甲酸　　HCOOH　　　　　methanoic acid
乙酸　　CH_3COOH　　　ethanoic acid
丙酸　　CH_3CH_2COOH　propanoic acid

- 习惯命名法

HCOOH　　　　　　formic acid
CH₃COOH　　　　　acetic acid
CH₃CH₂COOH　　　propionic acid

- Aromatic acid

The simplest aromatic acid is benzoic acid. *Ortho-hydroxy-benzoic acid* is known as salicylic acid, the basis for many *salicylate* drugs such as aspirin.

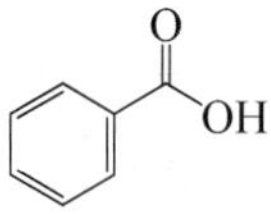

benzoic acid 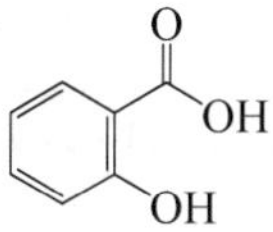　　　　salicylic acid 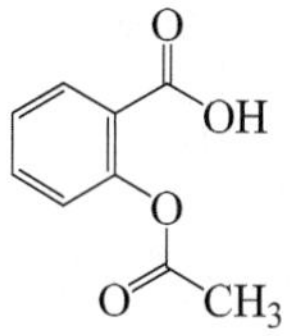　　　　acetylsalicylic acid(aspirin)

(7) 酯

① Esters are named like salts, giving them an **-ate** ending. The alcohol is named first, followed by the name of the acid modified to end in **-ate**. The **-ic** of the acid name is dropped and the letters **-ate** are added. Thus, acetic becomes acetate and ethanoic becomes ethanoate.

② The ester derived form methyl alcohol and acetic acid is called methyl acetate or methyl ethanoate (CH₃COOCH₃).

- 先写醇后写酸，去"ic"加"ate"

甲酸乙酯　　ethyl methanoate　　or　　ethyl formate
乙酸甲酯　　methyl acetate　　　or　　methyl ethanoate

Exercises

Give English name of the following compounds

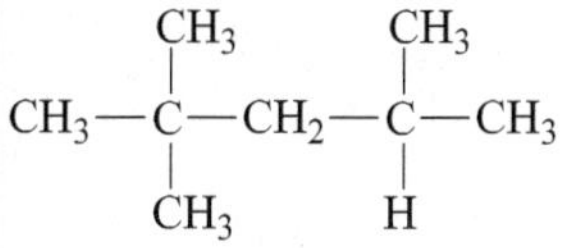 CH₃—C(CH₃)(CH₃)—CH₂—C(H)(CH₃)—CH₃	2,2-二甲基-4-乙基己烷	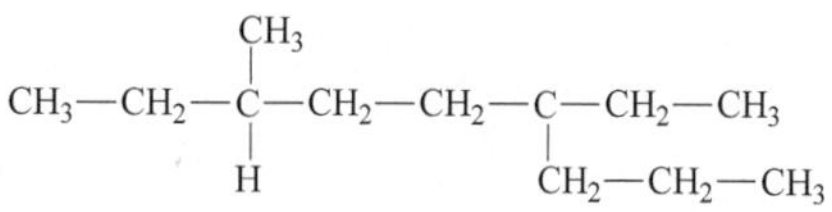 CH₃—CH₂—C(H)(CH₃)—CH₂—CH₂—C(CH₂CH₂CH₃)—CH₂—CH₃
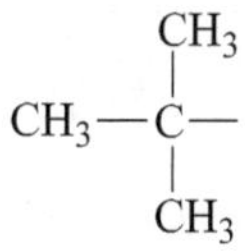 CH₃—C(CH₃)(CH₃)—CH₃	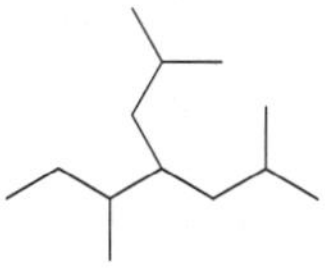	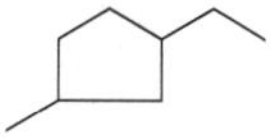
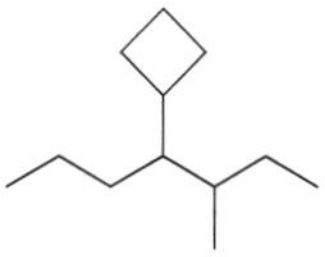	HC≡C—CH=CH—CH₃	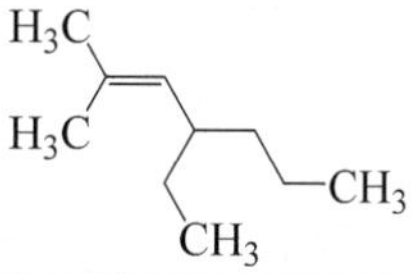

续表

(1,2-二甲基环己烯结构)	(甲苯结构)	(间乙基甲苯结构)
苯乙烯	苯基	苄基
甲醇	乙醇	丙烯醇
2,3-二甲基-2,3-丁二醇	叔丁醇	苯甲醇
苯酚	间三苯酚	CH_3OCH_3
甲乙醚	$CH_3C=CHCH_2CH_2CHO$ $\quad\mid$ $\quad CH_3$	甲醛
丁酮	乙酸	己二酸
苯甲酸	乙酸乙酯	乙酸甲酯

Part III

Reading and Comprehension of Scientific Articles

Unit 1 Chemical Technology and Engineering

Lesson 1 Chemical Industry

1. Origins of the chemical industry

It may be considered to have begun during the Industry Revolution, about 1800, and develop to provide chemicals for use by other industries. Examples are alkali for soapmaking, bleaching powder for cotton, and silica and sodium carbonate for glassmaking. It will be noted that these are all inorganic chemicals. The organic chemicals industry started in the 1860s with the exploitation of William Henry Perkin's discovery of the first synthetic dyestuff – mauve. At the start of the twentieth century the emphasis on research on the applied aspects of chemistry in German had paid off handsomely, and by 1914 had resulted in the German chemical industry having 75% of the world market in chemicals. This was based on the discovery of new dyestuffs plus the development of both the contact process for sulphuric acid and the Haber process for ammonia. The latter required a major technological breakthrough that of being able to carry out chemical reactions under conditions of very high pressure for the first time. The experience gained with this was to stand Germany in good stead, particularly with the rapidly increase demand for nitrogen-based compounds (ammonium salts for fertilizers and nitric acid for explosive manufacture) with the outbreak of World War I in 1914. This initiated profound changes which has continued during the inter-war years (1918-1939).

Since 1940 the chemical industry has grown at a remarkable rate, although this has slowed significantly recent years. The lion's share of this growth has been in the organic chemicals sector due to the development and growth of the petrochemicals area since 1950. The explosive growth in petrochemicals in the 1960s and 1970s was largely due to the enormous increase in demand for synthetic polymers such as polyethylene, polypropylene, nylon, polyesters and epoxy resins.

The chemical industry today is a very diverse sector of manufacturing industry, within which it plays a central role. It makes thousands of different chemicals which the general public only usually encounter as end or consumer products. These products are purchased because they have the required properties which make them suitable for some particular application, e.g., a non-stick coating for pans or a weedkiller. Thus chemicals are ultimately sold for the effects that they produce.

2. The need for chemical industry

Do we need chemical industry? Trying to answer this question will provide (ⅰ) an indication of the range of the chemical industry's activities, (ⅱ) its influence on our lives in everyday terms, and (ⅲ) how great is society's need for a chemical industry. Our approach in answering the question will consider the industry's contribution to meeting and satisfying our major needs. Clearly food (and drink) and health are paramount. Others which we shall consider in their turn are clothing and (briefly) shelter, leisure and transport.

① Food. The chemical industry makes a major contribution to food production in at least three ways. Firstly, by making an available large quantities of artificial fertilizers which are used to replace the elements (mainly nitrogen, phosphorus and potassium) which are removed as nutrients by the growing crops during modern intensive farming. Secondly, by manufacturing crop protection chemicals, i.e., pesticides, which markedly reduce the proportion of the crops consumed by pests. Thirdly, by producing verinary products which protect livestock from disease or cure their infections.

② Health. We are all aware of the major contribution which the pharmaceutical sector of the industry has made to help us keep healthy, e.g., by curing bacterial infections with antibiotics, and even extending life itself, e.g., β-blockers to lower blood pressure.

③ Clothing. The improvement in properties of modern synthetic fibers over the traditional clothing materials has been quite remarkable. Thus shirts, dresses and suits made from polyesters like terylene and polyamide like nylon are crease-resistant, machine-washable, and drip-dry or non-iron. They are also cheaper than natural materials.

Parallel developments in the discovery of modern synthetic dyes and the technology to "bond" them to the fiber has resulted in a tremendous increase in the variety of colors available to the fashion designer. Indeed they now span almost every color and hue of the visible spectrum. Indeed if a suitable shade is not available, structural modification of an existing dye to achieve this can readily be carried out, provided there is a satisfactory market for the product.

Other major advances in this sphere have been in color-fastness, i.e., resistance to the dyne being washed out when the garment is cleaned.

④ Shelter, leisure and transport. In terms of shelter the contribution of modern synthetic polymers has been substantial. Plastics are tending to replace traditional building materials like wood because they are lighter, mainternance-free (i.e., they are resistant to weathering and do not need painting). Other polymers, e.g., urea-formaldehyde and polyurethanes, are important insulating materials for reducing heat losses and hence reducing energy usage.

Plastics and polymers have made a considerable impact on leisure activities with applications ranging from all-weather artificial surfaces for athletic tracks, football pitches and tennis courts to nylon strings for racquets and items like golf balls and footballs made entirely from synthetic materials.

Likewise the chemical industry's contribution to transport over the years has led to major improvements. Thus development of improved additive like anti-oxidants and viscosity index improvers for engine oil has enabled routine servicing intervals to increase from 3000 to 6000 to 12000 miles. Research and development work has also resulted in improved lubricating oils and greases, and better brake fluids. Yet again the contribution of polymers and plastics has been very striking with the proportion of the total automobile derived form these materials—dashboard, steering wheel, seat padding and covering, etc. —now exceeding 40%.

So it is quite apparent even from a brief look at the chemical industry's contribution to meeting our major needs that life in the world would be very different without the products of the industry. Indeed the level of a country's development may be judged by the production level and sophistication of its chemical industry.

(Alan Heaton. The Chemical Industry. 2nd Edition. Blackie & Son Ltd, 1997.)

Words and Expressions

bleaching   adj. 漂白的; n. 漂白
dyestuff   n. 染料，[助剂] 着色剂
mauve   n. 淡紫色；苯胺紫；淡紫色染料；adj. 淡紫色的
ammonia   n. 氨
fertilizer   n. 肥料
emulsion   n. 乳剂，乳状液
vinyl   n. 乙烯基
fine chemical   n. 精细化学品

Exercises

1. Completing the following table, by listing the chemicals as many as you can

Some chemicals used in our daily life

Food		Shelter	
Health		Leisure	
Clothing		Transport	

2. Put the following into Chinese

carbonate	polypropylene	epoxy	vinyl
acetate	pharmaceutical	spectrum	fromaldehyde
silica	ammonium	polyester	the lion's share

3. Put the following into English

Na	K	P	氨
聚合物	聚乙烯	氯化物	黏度
烃	催化剂	炼油厂	添加剂

Lesson 2 Research and Development

Research and development, or R&D as it is commonly referred to, is an activity which is carried out by all sectors of manufacturing industry but its extent varies considerably, as we will see shortly. Although the distinction between research and development is not always clear-cut, and there is often considerable overlap, we will attempt to separate them. In simple terms research can be thought of as the activity which produces new ideas and knowledge whereas development is putting those ideas into practice as new processes and products. To illustrate this with an example, predicting the structure of a new molecule which would have a specific biological activity, and synthesizing it could be seen as research, whereas testing it and developing it to the point where it could be marketed as a new drug could be described as the development part.

1. Fundamental research and applied research

In industry the primary reason for carrying out R&D is economic and is to strengthen and improve the company's position and profitability. The purpose of R&D is to generate and to provide information and knowledge to reduce uncertainty, to solve problems and to provide better data on which management can base decisions. Specific projects cover a wide range of activities and time scales, from a few months to 20 years.

We can pick out a number of areas of R&D activity in the following paragraphs but if we were to start with those which were to spring to the mind of the academic, rather than the industrial chemist, then these would be basic, fundamental (background) or exploratory research and the synthesis of new compounds. This is also labeled "blue-sky" research.

Fundamental research is typically associated with university research. It may be carried out for its own intrinsic interest and it will add to the total knowledge base but no immediate applications of it in the "real world" will be apparent. Note that it will provide a valuable training in defining and solving problems, i.e., research methodology for the research student who carries it out, under supervision. However, later "spin offs" from such work can lead to useful applications. Thus physicists claim that but for the study and development of quantum theory we might not have had computers and nuclear power. However, to take a specifically chemical example, general studies on a broad area such as hydrocarbon oxidation might provide information which would be useful in more specific areas such as cyclohexane oxidation for the production of nylon intermediates.

Aspects of synthesis could involve either developing new, more specific reagents for controlling particular functional group interconversions, i.e., developing synthetic methodology or complete synthesis of an entirely new molecule which is biologically active. Although the former is clearly fundamental the latter encompasses both this and applied aspects. This term "applied" has traditionally been more associated with research carried out in industrial laboratories, since this is more focused or targeted. It is a consequence of the work being business driven.

Note, however, that there has been a major change in recent years as academic institutions have increasingly turned to industry for research funding, with the result that much more of their research effort is now devoted to more applied research. Even so, in academia the emphasis generally is very much on the research rather than the development.

2. Types of industrial research and development

The applied or more targeted type of research and development commonly carried out in industry can be of several types and we will briefly consider each. They are: product development, process development, process improvement and applications development. Even under these headings there are a multitude of aspects so only a typical example can be quoted in each case.

① Product development. Product development includes not only the discovery and development of a new drug but also, for example, providing a new longer-acting anti-oxidant additive to an automobile engine oil. Developments such as this have enabled servicing intervals to increase during the last decade from 3000 to 6000 to 9000 and now to 12000 miles. Note that most purchasers of chemicals acquire them for the effects that they produce, i.e., a specific use. Teflon^{TM}, or polytetrafluoroethylene(PTEE), may be purchased because it imparts a non-stick surface to cooking pots and pans, thereby making them easier to clean.

② Process development. Process development covers not only developing a manufacturing process for an entirely new product but also a new process or route for an existing product. The push for the latter may originate for one or more of the following reasons: availability of new technology, change in the availability and/or cost of raw materials. Manufacture of vinyl chloride monomer is an example of this. Its manufacturing route has been changed several times owing to changing economics, technology and raw materials. Another stimulus is marked increase in demand and hence sales volume which can have a major effect on the economics of the process. The early days of penicillin manufacture afford a good example of this.

③ Process improvement. Process improvement relates to processes which are already operating. It may be a problem that has arisen and stopped production. In this situation there is a lot of pressure to find a solution as soon as possible so that production can restart, since "down time" costs money.

④ Applications development. Clearly the discovery of new applications or uses for a product can increase or prolong its profitability. Not only does this generate more income but the resulting increased scale of production can lead to lower unit costs and increased profit. An example is PVC whose early use included records and plastic raincoats. Applications which came later included plastic bags and particularly engineering uses in pipes and guttering.

(Heaton C A. An Introduction to Industrial Chemistry. 2nd Edition. Blackie &Son Ltd, 1997.)

Words and Expressions

clear-cut   adj. 明确的，鲜明的
blue-sky   adj. 纯理论的
intrinsic   adj. 本质的，固有的
spin off   n. 伴随（附带）的结果
impart   vt. 给予（尤指抽象事物），传授，告知，透露
penicillin   n. 盘尼西林（青霉素）
strain   n. 菌株（种）
mould   n. 霉菌
semi-technical   adj. 半工业化的

pilot-plant   n. 中间试验工厂
envisage   vt. 正视，面对，想象
chloralkali   n. 氯碱

Exercises

1. Put the following into Chinese

quantum	strain	mould	phenol
sulphate	carbide	foul	scrub
semi-technical	fermenter	CFC	refrigerant

2. Put the following into English

试剂	单体	丙酮	脉动
乙炔	硫	盐酸	停车时间
杂质	反应器	（使）优化	纯度

Lesson 3 Typical Activities of Chemical Engineers

The classical role of the chemical engineer is to take the discoveries made by the chemist in the laboratory and develop them into money-making, commercial-scale chemical process. The chemist works in test tubes and Parr bombs with very small quantities of reactants and products (e.g.,100ml), usually running "batch", constant-temperature experiments. Reactants are placed in a small container in a constant temperature bath. A catalyst is added and the reactions proceed with time. Samples are taken at appropriate intervals to follow the consumption of the reactants and the production of products as time progresses.

By contrast, the chemical engineers typically work with much larger quantities of material and with very large (and expensive) equipment. Distillation columns can be over 100 feet high and 10 to 30 feet in diameter. The capital investment for one process unit in a chemical plant may exceed $100 million!

The chemical engineer is often involved in "scaling up" a chemist-developed small-scale reactor and separation system to a very large commercial plant. The chemical engineer must work closely with the chemist in order to understand thoroughly the chemistry involved in the process and to make sure that the chemists get the reaction kinetic data and the physical property data needed to design, operate, and optimize the process. This is why the chemical engineering curriculum contains so many chemistry courses.

The chemical engineer must also work closely with mechanical, electrical, civil, and metallurgical engineers in order to design and operate the physical equipment in a plant—the reactors, tanks, distillation columns, heat exchangers, pumps, compressors, control and instrumentation devices, and so on.

To commercialize the laboratory chemistry, the chemical engineer is involved in development, design, construction, operation, scales and research. Let's describe each of these functions briefly.

1. Development

Development is the intermediate step required in passing from a laboratory-size process to a commercial-size process. The "pilot-plant" process involved in development might involve reactors that are five gallons in capacity and distillation columns that are three inches in diameter. Development is usually part of the commercialization of a chemical process because the scale-up problem is very different one. Jumping directly from test tubes to 10000-gallon reactors can be a tricky and sometimes dangerous endeavor.

The chemical engineer works with the chemist and a team of other engineers to design, construct and operate the pilot plant.

Once the pilot plant is operational, performance and optimization data can be obtained in order to evaluate the process from an economic point of view. The profitability is assessed at each stage of the development of the process. If it appears that not enough money will be made to justify the capital investment, the project will be stopped.

The pilot plant offers the opportunity to evaluate materials of construction, measurement techniques, and process control strategies. The experimental findings in the pilot plant can be used to improve the design of the full-scale plant.

2. Design

Based on the experience and data obtained in the laboratory and the pilot plant, a team of engineers is assembled to design the commercial plant. The chemical engineer's job is to specify all process flow rates and conditions, equipment types and sizes, materials of construction, process configurations, control system, safety systems, environmental protection systems, and other relevant specifications. It is an enormous responsibility.

The product of the design stage is a lot of paper.

① *Flow Sheets* are diagrams showing all the equipment schematically, with all streams labled and their conditions specified (flow rate, temperature, pressure, composition, ciscosity, density, etc.)

② *P and I (Piping and Instrumentation) Drawings* are drawings showing all pieces of equipment (including sizes, nozzle locations, and materials), all piping (including sizes, materials and valves), all instrumentation (including locations and types of sensors, control valves, and controllers), and all safety systems (including safety valve and rupture disk locations and sizes, flare lines, and safe operating conditions).

③ *Equipment Specification Sheets* are sheets of detailed information on all the equipment precise dimentions, performance criteria, materials of construction, corrosion allowances, operating temperatures, and pressures, maximum and minimum flow rates, and the like. These are sent to equipment manufacturers for price bids and then for building the equipment.

3. Construction

After the equipment manufacturers have built the individual pieces of equipment, the pieces are shipped to the plant site (sometimes a challenging job of logistics, particularly for large vessels like distillationcolumns). The construction phase is the assembling of all the components into a complete plant.

This is usually a very exciting and rewarding time for most engineers. Your are seeing your ideas being translated from paper into reality. Steel and concrete replace sketches and diagrams.

The engineers are usually on shift work during the startup period. There is a lot to learn in a short time period. Once the plant has been successfully operated at its rated performance, it is turned over to the operating or manufacturing department for routine production of products.

4. Manufacturing

Chemical engineers occupy a central position in manufacturing. Plant technical service groups are responsible for the technical aspects of running an efficient and safe plant. They run capacity and performance tests on the plant to determine where the bottlenecks are in the equipment, and then design modifications and additions to remove these bottlenecks.

Chemical engineers study ways to reduce operating cost by saving energy, cutting raw material consumption, and reducing production of off-specification products that require reprocessing. They study ways to improve product quality and reduce environmental pollution of both air and water.

5. Technical Sales

Many chemical engineers find stimulating and profitable careers in technical sales. As with other sales positions, the work involves calling on customers, making recommendations on particular products to fill customer's needs, and being sure that orders are handled smoothly. The sales engineer is the company's representative and must know the company's product line well.

6. Research

Chemical engineers are engaged in many types of research. They work with chemist in developing new or improved product. They develop new and improved engineering methods. They work on improved sensors for on-line physical property measurements. They study alternative process configurations and equipments.

(William L L. Chemical Process Analysis. Prentice Hall, 1988.)

Exercises

1. Put the following into Chinese

reactant	distillation	compressor	pilot-plant
specification	flow sheet	nozzle	corrosion
sensor	strophy	on-line	commission

2. Put the following into English

间歇的	反应器	放大	热交换器
创新	术语	阀	流程图
梯度	组成	杂质	模拟

Lesson 4 Sources of Chemicals

The number and diversity of chemical compounds is remarkable: over ten million are now known. Even this vast number pales into insignificance when compared to the number of carbon compounds which is theoretically possible. This is a consequence of catenation, i.e., formation of very long chains of carbon atoms due to the relatively strong carbon-carbon covalent bonds, and isomerism. Most of these compounds are merely laboratory curiosities or are only of academic interest. However, of the remainder there are probably several thousands which are of commercial and practical interest. It might therefore be expected that there would be a large number of sources of these chemicals. Although this is true for inorganic chemicals, surprisingly most organic chemicals can originate from a single source such as crude oil (petroleum).

1. Inorganic chemicals

Table 3 Major sources of inorganic chemicals

source	Examples of uses
phosphate rock	fertilizers, detergents
salt	chlorine, alkali production
limestone	soda ash, lime, calcium carbide
sulphur	sulphuric acid production
potassium compounds	caustic potash, fertilizers
bauxite	aluminum salts
sodium carbonate	caustic soda, cleaning formulations
titanium compounds	titanium dioxide pigments, lightweight alloys
magnesite	magnesium salts
borates	borax, boric acid, glazes
fluorite	aluminum fluoride, organofluorine compounds

Since the term "inorganic chemical" covers compounds of all the elements other than carbon, the diversity of origins is not surprising (Table 3). Some of the more important sources are metallic ores (for important metals like iron and aluminum), and salt or brine (for chlorine, sodium, sodium hydroxide and sodium carbonate). In all these cases at least two different elements are combined together chemically in the form of a stable compound. If therefore the individual element or elements, say the metal, are required then the extraction process must involve chemical treatment in addition to any separation methods of a purely physical nature. Metal ores, or minerals, rarely occur on their own in a pure form and therefore a first step in their processing is usually the separation from unwanted solids, such as clay or sand. Crushing and grinding of the solids followed by sieving may achieve some physical separation because of the differing particle size. The next stage depends on the nature and properties of the required ore. For example, iron-bearing ores can often be separated by utilizing their magnetic properties in a magnetic separator. Froth flotation is another widely used technique in which the desired ore, in a fine particulate form, is separated from other solids by a difference in their ability to be wetted by an aqueous solution. Surface active (anti-wetting) agents are added to the solution, and these are typically molecules having a non-polar part, e.g., a long hydrocarbon chain, with a polar part such as an amino group at one end. This polar grouping attracts the ore, forming a loose bond. The hydrocarbon grouping now repels the water, thus preventing the ore being wetted, and it therefore floats. Other solids, in contrast, are readily wetted and therefore sink in the aqueous solution. Stirring or bubbling the liquid to give a froth considerably aids the "floating" of the agent-coated ore which the overflows from this tank into a collecting vessel, where it can be recovered. The key to success is clearly in the choice of a highly specific surface-active agent for the ore in question.

2. Organic compounds

In contrast to inorganic chemicals which, as we have already seen, are derived from many different sources, the multitude of commercially important organic compounds are essentially derived from a single source. Nowadays in excess of 99% (by tonnage) of all organic chemicals are obtained from crude oil (petroleum) and natural gas via petrochemical processes. This is a very interesting situation—one which has changed over the years and will change again in the future—because technically these same chemicals could be obtained from other raw materials or sources. Thus aliphatic compounds, in particular, may be produced via ethanol, which is obtained by fermentation of carbohydrates. Aromatic compounds on the other hand are isolated from coal-tar, which is a by-product in the carbonization of coal.

Organic chemicals from carbohydrates (biomass): The main constituents of plants are carbohydrates which comprise the structural parts of the plant. They are polysaccharides such as cellulose and starch. Starch occurs in the plant kingdom in large quantities in foods such as cereals, rice and potatoes, cellulose is the primary substance form which the walls of plant cells are constructed and therefore occurs very widely and may be obtained from wood, cotton, etc. Thus, not noly is the potential for chemicals considerable, but the feedstock is renewable.

The major route from biomass to chemicals is via fermentation processes.

Fermentation processes utilize single-cell micro-organisms typically yeasts, fungi, bacteria or moulds to produce particular chemicals. Some of these processes have been used in the domestic situation for many thousands of years, the best-known example being fermentation of grains to produce alcoholic beverages. Indeed up until about 1950 this was the most popular route to aliphatic organic chemicals, since the ethanol produced could be dehydrated to give ethylene, which is the key intermediate for the synthesis of a whole range of aliphatic compounds.

Disadvantages reflected in this can be divided into two parts: (i) raw materials, (ii) the fermentation process. Raw-material costs are higher than that of crude oil. Major disadvantages are, firstly the time scale, and secondly, the product is obtained as a dilute aqueous solution. The separation and purification costs are very high. Since micro-organism is a living system, little variation in process is permitted.

On the other hand particular advantages of fermentation methods are that they are very selective and some chemicals which are structurally very complex, and therefore extremely difficult to synthesize, and/or require a multi-stage synthesis, are easily made. Notable examples are various antibiotics, penicillins, cephalosporins, streptomycins.

(Alan Heaton.The Chemical Industry. 2nd Edition.Blackie & Son Ltd, 1997.)

Words and Expressions

pale   adj. 苍白的；无力的；vt. 使失色；使变苍白；vi. 失色；变苍白
catenation   n. 耦合，连接
isomerism   n. 同分异构（现象）
detergent   n. 清洁剂，去垢剂
bauxite   n. 铁铝氧石，铝土矿
pigment   n. 色素；颜料；vt. 给……着色；vi. 呈现颜色
glaze   n. 釉面；光滑面；vt. 给陶（瓷）器上釉

Exercises

1. Put the following into Chinese

covalent   isomerism   froth flotation   borate

fluoride	amino	hydrolysis	ester
naphthene	naphtha	enzymic	xylene

2. Put the following into English

氢氧化物	脂肪族的	芳香族的	甲烷
酯	不饱和的	烯烃	烷烃

Lesson 5 Basic Chemicals

We can divide the various sectors of the chemical industry into these two types: the high-volume sector and low-volume sector. In the high-volume sector, individual chemicals are typically produced on the tens to hundreds of thousands of tonnes per annum scale. As a result, the plants used are dedicated to the single product, operate in a continuous manner and are highly automated, including computer control. Sectors categorized as high-volume are sulphuric acid, phosphorus-containing compounds, nitrogen-containing compounds, chlor-alkali and related compounds, plus petrochemicals and commodity polymers such as polythene. With the exception of the latter, they are key intermediates, or basic chemicals, which are feedstock for the production of a wide range of other chemicals, many of which are also required in large quantities.

In contrast, low-volume sectors are largely involved in fine-chemicals manufacture, and individual products are produced only on the tens of tonnes to possibly a few thousands tonnes scale. However, they have a very high value per unit weight, in contrast to high-volume products. Fine-chemicals are usually produced in plants operating in a batch manner and the plants may be multiproduct ones. Sectors are agrochemicals, dyestuffs, pharmaceutical, and speciality polymers such as PEEK.

Basic chemicals are the orphans of the chemical industry. They are not glamorous, like drugs and are sometimes not very profitable (and at the very least the profits come in unpredictable cycles of boom and bust). They are not seen or used directly by the general public and so their importance is not often understood. Even within the industry their importance is often insufficiently appreciated. Without them, however, the rest of the industry could not exist.

Basic chemicals occupy the middle ground between raw materials and end products. One distinguishing feature of basic chemicals is the scale on which they are manufactured; everything from really big to absolutely enormous. Fig.2-1❶ shows the top 25 chemicals in the USA market by volume in 1993, just to give a feel for the sort of chemicals and volumes concerned. Basic chemicals are typically manufactured in plants that produce hundreds of thousands of tones of products per year. A plant that produces 100,000 tonnes

---

❶ 此为原文配图，与本书关联不大，故不列入。——编者注

per year will produce about 12.5 tonnes per hour. Another distinguishing and important feature is their price: most of them are fairly cheap.

The job of the basic chemical industry is to find economical ways of turning raw materials into useful intermediates. There is little leeway for any company to charge premium prices for its products, so the company that makes the cheapest cost will probably be the most profitable. This situation means that companies must always be on their toes looking for new and more economical ways of making and transforming their raw materials.

Many basic chemicals are the products of oil refining, while parts of the industry—the sulphur, nitrogen, phosphorus and chloro-alkali industry—put elements other than carbon and hydrogen into chemicals. In combination, these products and the basic products of the petrochemical industry can be combined to produce the myriad of important chemicals that feed the rest of the chemical industry.

The basic chemicals industry is now facing one of the biggest challenges in its history. The main consumer of the industry's key products—the agriculture industry—has stopped growing and is severely cutting back demand for fertilizers. Western farmers have been producing too much food and governments have been cutting subsidies, with the result that less land is being farmed and less fertilizers used. Environmental concerns about the effects of excessive fertilizer run-off have also reduced demand for fertilizers.

Products such as chlorinated compounds have come under threat from environmentalists. Some will be banned under the Montreal Protocal, but others are not harmful and may survive under environmentalist pressures. The industry can no longer rely on long-term growth in demand.

The industry may well see increased consolidation as companies swap plants to achieve better economies of scale or better market position in specific products. This could leave an industry with far fewer players but with a better balance of supply and demand and better profitability. The industry will move more to serving the rest of the chemical industry and less to serving the farming industry.

Another threat is the perceived environmental messiness of many large-scale processes. Despite the relative efficiency of many big plants, the industry has a long way to go to achieve the best environmental standards possibly. The drive to increased recycling and the ideal of emission-free plants will be a major factor influencing the development of the industry in the next decade.

Technical developments will not stop. There will be increasing emphasis on plants and processes that do not pollute. Companies will compete on efficiency—those able to produce the best quality products at the cheapest price will prosper. This will require companies to keep investing in technical improvements. New ways of bringing basic chemicals together to form useful intermediates will be found.

There is still much to do in the basic chemicals industry.

(Alan Heaton.The Chemical Industry. 2nd Edition.Blackie & Son Ltd, 1997.)

Words and Expressions

dedicate   vt. 把（时间、力量等）用在……（to），奉献
commodity   n. 商品，货物，日用品
speciality   n. 特制品，特殊产品，专业化
premium   n. 奖励，额外费用，奖金
on one's toes   准备行动的
run-off   n. 流出，流泻，径流
emission-free   adj. 无排放的，零排放的

Exercises

1. Summarize the information on the two sectors of chemical industry by completing the table below

Information	High-volume sector	Low-volume sector
Production scale		
Products/a plant		
Operation manner		
Price or profit		
Usage		
Challenges		
Products in the sector		

2. Put the following into Chinese

commodity	ether	speciality	end-product
on one's toes	elastomer	hydrate	plasticizer
sulphonate	formulate	metallurgy	phosphoric acid

3. Put the following into English

烷烃	芳基	乙基	丁基
离子	乙醇	甲醇	醋酸
均相的	系数	摩擦	无排放的

Lesson 6 Ammonia

Dinitrogen makes up more than three-quarters of the air we breathe, but it is not readily available for further chemical use. Biological transformation of nitrogen into useful chemicals is embarrassing for the chemical industry, since all the effort of all the industry's technologists has been unable to find an easy alternative to this. Leguminous plants can take nitrogen from the air and convert it into ammonia and ammonia-containing products at atmospheric pressure ad ambient temperature; despite a hundred years of effort, the chemical industry still needs high temperatures and pressures of hundreds of atmospheres to do the same job. Indeed, until the invention of the Haber process, all nitrogen-containing chemicals came from mineral sources ultimately derived from biological activity.

Essentially all the nitrogen in manufactured chemicals comes from ammonia derived from the Haber-based process. So much ammonia is made (more molecules than any other compound because it is a light molecule, although greater weights of other products are produced), and so energy-intensive is the process, that ammonia production alone was estimated to use 3% of the world's energy supply in the mid-1980s.

The Haber Process for Ammonia Synthesis

1. Introduction

All methods for making ammonia are basically fine-tuned versions of the process developed by Haber, Nernst and Bosch in Germany just before the First World War.

$$N_2+3H_2 \rightleftharpoons 2NH_3$$

In principle the reaction between hydrogen and nitrogen is easy; it is exothermic and the equilibrium lies to the right at low temperatures. Unfortunately, nature has bestowed dinitrogen with an inconveniently strong triple bond, enabling the molecule to thumb its nose at thermodynamics. In scientific terms the molecule is kinetically inert, and rather severe reaction conditions are necessary to get reactions to proceed at a respectable rate. A major source of "fix" (meaning, paradoxically, "usefully reactive") nitrogen in nature is lighting, where the intense heat is sufficient to create nitrogen oxides from nitrogen and oxygen.

To get a respectable yield of ammonia in a chemical plant we need to use a catalyst. What Haber discovered—and it won his a Nobel Prize—was that some iron compounds were acceptable catalysts. Even with such catalysts extreme pressures (up to 600 atmospheres in early processes) and temperatures (perhaps 400℃) are necessary.

Pressure drives the equilibrium forward, as four molecules of gas are being transformed into two. Higher temperatures, however, drive the equilibrium the wrong way, though they do make the reaction faster, chosen conditions must be a compromise that gives an acceptable conversion at a reasonable speed. The precise choice will depend on other economic factors and the details of the catalyst. Modern plants have tended to operate at lower pressures and higher temperatures (recycling unconverted material) than the nearer-ideal early plants, since the capital and energy costs have become more significant.

Biological fixation also uses a catalyst which contains molybdenum (or vanadium) and iron embedded in a very large protein, the detailed structure of which eluded chemists until late 1992. How it works is still not understood in detail.

2. Raw materials

The process requires several inputs: energy, nitrogen and hydrogen. Nitrogen is easy to extract from air, but hydrogen is another problem. Originally it was derived from coal via coke which can be used as a raw material (basically a source of carbon) in steam reforming, where steam is reacted with carbon to give hydrogen, carbon monoxide and carbon dioxide. Now natural gas (mainly methane) is used instead, though other hydrocarbons from oil can also be used. Ammonia plants always include hydrogen-producing plants linked directly to the production of ammonia.

Prior to reforming reactions, sulphur-containing compounds must be removed from the hydrocarbon feedstock as they poison both the reforming catalysts and the Haber catalysts. The first desulphurisation stage involves a cobalt-molybdenum catalyst, which hydrogenates all sulphur-containing compounds to hydrogen sulfide. This can then be removed by reaction with zinc oxide (to give zinc sulfide and water).

The major reforming reactions are typified by the following reactions of methane (which occur over nickel-based catalysts at about 750 ℃):

$$CH_4 + H_2O \longrightarrow CO + 3H_2$$
$$CH_4 + 2H_2O \longrightarrow CO_2 + 4H_2$$

Other hydrocarbons undergo similar reactions.

In the secondary reformers, air is injected into the gas steam at about 1100 ℃. In addition to the other reactions occurring, the oxygen reacts with hydrogen to give water, leaving a mixture with close to the ideal 3:1 ratio of hydrogen to nitrogen with no contaminating oxygen. Further reactions, however, are necessary to convert more of the carbon monoxide into hydrogen and carbon dioxide via the shift reaction:

$$CO + H_2O \longrightarrow CO_2 + H_2$$

This reaction is carried out at lower temperatures and in two stages (400 ℃ with an iron catalyst and 220 ℃ with a copper catalyst) to ensure that conversion is as complete as possible.

In the next stage, carbon dioxide must be removed from the gas mixture, and this is accomplished by reacting the acidic gas with an alkaline solution such as potassium hydroxide and/or mono- or di-ethanolamine.

By this stage there is still too much contamination of the hydrogen-nitrogen mixture by carbon monoxide (which poisons the Haber catalysts), and another step is needed to get amount of CO down to ppm levels. This step is called methanation and involves the reaction of CO and hydrogen to give methane (i.e., the reverse of some of the reforming steps). The reaction operates at about 325℃ and uses a nickel catalyst.

Now the synthesis gas mixture is ready to go into a Haber reaction.

3. Ammonia production

The common features of all the different varieties of ammonia plants are that the synthesis gas mixture is heated, compressed and passed into a reactor containing a catalyst. The essential equation for the reaction is simple:

$$N_2 + 3H_2 \rightleftharpoons 2NH_3$$

What industry needs to achieve in the process is an acceptable combination of reaction speed and reaction yield. Different compromises have been sought at different times and in different economic circumstances. Early plants plumped for very high pressure, but many of the most modern plants have accepted much lower one-pass yields at low pressures and have also opted for lower temperatures to conserve energy. In order to ensure the maximum yield in the reactor the synthesis gas is usually cooled as it reaches the equilibrium. This can be done by the use of heat exchangers or by the injection of cool gas into the reactors at an appropriate point. Since the reaction is exothermic the heat must be carefully controlled in this way to achieve good yields.

The output from the Haber stage will consist of a mixture of ammonia and synthesis gas, so the next stage needs to be the separation of the two so that the synthesis gas can be recycled. This is normally accomplished by condensing the ammonia (which is a good deal less volatile than the other components, ammonia boils at about $-40℃$).

4. Uses of ammonia

The major use of ammonia is not for the production of nitrogen-containing chemicals for further industry use, but for fertilizers such as urea or ammonium nitrates and phosphates. Fertilizers consume 80% of all the ammonia produced. In the USA in 1991, for example, the following ammonia-derived products were consumed, mostly for fertilizers (amounts in millions of tones): urea (4.2); ammonium sulphate (2.2); ammonium nitrate (2.6); diammonium hydrogen phosphate (13.5).

Chemical uses of ammonia are varied. The Solvay process for the manufacture of soda ash uses ammonia, though it does not appear in the final product since it is recycled. A wide variety of processes take in ammonia directly, including the production of cyanides and aromatic nitrogen-containing compounds such as pyridine. The nitrogen in many polymers (such as nylon or acrylics) can be traced back to ammonia, often via nitriles or hydrogen cyanide. Most other processes use nitric acid or salts derived from it as their source of nitrogen. Ammonium nitrate, used as a nitrogen-rich fertilizer, also finds a major use as a bulk explosive.

(Heaton C A. An Introduction to Industrial Chemistry. 2nd Edition. Blackie &Son Ltd, 1997.)

Words and Expressions

dinitrogen   n. 分子氮，二氮
leguminous   adj. 豆科的，似豆科植物的
ambient   adj. 周围的，包围着的
bestow   vt. 把……赠予（给）
thumb one's noise (at)   （对……）做蔑视的手势
thermodynamics   n. 热力学
paradoxically   adv. 似是而非地，自相矛盾地，荒谬地
molybdenum   n. 钼 Mo
embed   vt. 把……嵌入，栽种
elude   vt. 使困惑，难倒
cobalt   n. 钴
hydrogenate   vt. 使与氢化合，使氢化
secondary reformer   二段（次）转化炉
ethanolamine   n. 乙醇胺
methanation   n. 甲烷化作用
plump   vi. 投票赞成，坚决拥护（for）
one-pass   单程，非循环过程
opt   vi. 选择，挑选（for, between）
acrylic   adj. 聚丙烯的，丙烯酸（衍生物）的
nitrile   n. 腈

Exercises

1. Put the following into Chinese

soda ash	refractory	silicate	chromatography
mercury	alkaline	desulphurisation	membrane
anode	cathode	contaminate	inert

2. Put the following into English

电解	分解	复分解	还原
沉淀	结晶	过滤	吸收
溶解度	溶度积	平衡	放热的

Lesson 7　What is Chemical Engineering?

In a wider sense, *engineering* may be defined as a scientific presentation of the techniques and facilities used in a particular industry. For example, mechanical engineering refers to the techniques and facilities employed to make machines. It is predominantly based on mechanical forces which are used to change the appearance and/or physical properties of the materials being worked, while their chemical properties are left unchanged. Chemical engineering encompasses the chemical processing of raw materials, based on chemical and physico-chemical phenomena of high complexity.

Thus, chemical engineering is that branch of engineering which is concerned with the study of the design, manufacture, and operation of plant and machinery in industrial chemical processes.

Chemical engineering is the application of science, mathematics and economics to the process of converting raw materials or chemicals into more useful or valuable forms. Chemical engineering largely involves the design and maintenance of chemical processes for large-scale manufacture. Chemical engineers in this branch are usually employed under the title of process engineer.

The individual processes used by chemical engineers (e.g. distillation or chlorination) are called unit operations and consist of chemical reaction, mass-, heat- and momentum-transfer operations. Unit operations are grouped together in various configurations for the purpose of chemical synthesis and/or chemical separation.

1. A brief history outline

Chemical engineering is the profession concerned with the creative application of the scientific principles underlying the transport of mass, energy and momentum, and the physical and chemical change of matter. The broad implications of this definition have been justified over the past few decades by the kinds of problems that chemical engineers have solved, though the profession has devoted its attention in the main to the chemical process industries. As a result chemical engineers have been defined more traditionally as those applied scientists trained to deal with the research, development, design and operation problems of the chemicals, petroleum and related industries.

Chemical engineering developed as a distinct discipline during the twentieth century in answer to the needs of a chemical industry no longer able to operate efficiently with manufacturing processes which in many cases were simply larger scale versions of laboratory equipment. Thus, the primary emphasis in the profession was initially devoted to the general subject of how to use the results of laboratory experiments to design process equipment capable of meeting industrial production rates. This led naturally to the characterization of design procedures in terms of the unit operations, those elements common to many different processes. The basic unit operations include fluid flow, heat exchange, distillation, extraction, etc. A typical manufacturing process will be made up of combinations of the unit operations. Hence, skill in designing each of the units at a production scale would provide the means of designing the entire process.

The unit operations concept dominated chemical engineering education and practice until the mid-1950s, when a movement away from this equipment-oriented philosophy toward and *engineering science* approach began. This approach holds that the unifying concept is not specific processing operations, but rather the understanding of the fundamental phenomena of mass, energy and momentum transport that are common to all of the unit operations, and it is argued that concentration on unit operations obscures the similarity of many operations at a fundamental level.

Although there is no real conflict between the goals of the unit operations and engineering science approaches, the latter has tended to emphasize mathematical skill and to de-emphasize the design aspects of engineering education. One essential skill in reaching this goal is the ability to express engineering problems meaningfully in precise quantitative terms. Only in this way can the chemical engineer correctly formulate, interpret, and use fundamental experiments and physical principles in real world applications outside of the laboratory.

2. Related fields and topics

The modern discipline of chemical engineering enables much more than just process engineering. This often involves using chemical knowledge to create better materials and products that are useful in today's world. Chemical engineers are now engaged in development and production of diverse, high-value products, as well as in basic chemicals production. These products include specialty chemicals and high performance materials needed for aerospace, automotive, biomedical, electronic, environmental and military applications. Examples include ultra-strong fibers, fabrics, adhesives and composites for vehicles, bio-compatible materials for implants and prosthetics, gels for medical applications, pharmaceuticals, and films with special dielectric, optical or spectroscopic properties for opto-electronic devices. Additionally, chemical engineering is often intertwined with biology and biomedical engineering. Many chemical engineers work on biological projects such as understanding biopolymers (proteins) and mapping the human genome.

Today, the field of chemical engineering is a diverse one, covering areas from biotechnology and nanotechnology to mineral processing.

- Biochemical/ Biomedical engineering
- Biotechnology
- Environment

- Thermodynamics
- Transport phenomena (fluid dynamics, heat transfer, mass transfer)
- Materials science (ceramics and polymers)
- Nanotechnology
- Particle technology
- Chemical reactor
- Separation processes (see also: Separation of mixture)
  ◇ Membrane processes
  ◇ Distillation processes
  ◇ Crystallization processes
- Process control, process design, process modeling

Three primary physical laws underlying chemical engineering design are Conservation of Mass, Conservation of Momentum and Conservation of Energy. The movement of mass and energy around a chemical process are evaluated using mass and energy balances which apply these laws to whole plants, unit operations or discrete parts of equipment. In doing so, Chemical engineers use principles of thermodynamics, reaction kinetics and transport phenomena. The task of performing these balances is now aided by process simulators, which are complex software models that can solve mass and energy balances and usually have built-in modules to simulate a variety of common unit operations.

(Kutepov A M. Basic Chemical Engineering, Mir Publishers, 1988; William H. Brock. The Frontana History of Chemistry. Fontana Press, 1992.)

Words and Expressions

accessory   n. 配件，附件
encyclopedia   n. 百科全书
generalization   n. 概括，普遍化，一般化
with the advent of   随着……的到来（出现）
encapsulate [in'kæpsəleit]   vt. 压缩，节约，用胶囊包起来，将……封进内部

Exercises

Translate the following into Chinese

① Chemical engineering deals with the design, construction, and operation of plants and machinery for making such products as acids, dyes, drugs, plastics, and synthetic rubber by adapting the chemical reactions discovered by the laboratory chemist to large-scale production. The chemical engineer must be familiar with both chemistry and mechanical engineering.

② Following is an example that illustrates the process engineering part of chemical engineering:

The difference between chemical engineering and chemistry can be illustrated by considering the example of producing orange juice. A chemist working in the laboratory investigates and discovers a multitude of pathways to extract the juice of an orange. The simplest mechanism found is to cut the orange in half and squeeze the orange using a manual juicer. A more complicated approach found is to peel and then crush the orange and collect the juice. A company then commissions a chemical engineer to design a plant to manufacture several thousand tons of orange juice per year. The chemical engineer investigates all the available methods for making orange juice and evaluates them according to their economical viability. Even though the manual juicing method is simple, it is not economical to employ thousands of people to manually juice oranges. Thus another, cheaper method is used (possibly the "peel and crush" technique). The easiest method of manufacture on a laboratory bench will not necessarily be the most economical method for a manufacturing plant.

Lesson 8　Unit Operation in Chemical Engineering

Chemical processes may consist of widely varying sequences of steps, the principles of which are independent of the material being operated upon and of other characteristics of the particular system. In the design of a process, each step to be used can be studied individually if the steps are recognized. Some of the steps are chemical reactions, whereas others are physical changes. The versatility of chemical engineering originates in training to the practice of breaking up a complex process into individual physical steps, called unit operation, and into the chemical reactions. The unit-operations concept in chemical engineering is based on the philosophy that the widely varying sequences of steps can be reduced to simple operations or reactions, which are identical in fundamentals regardless of the material being processed. This principle, which became obvious to the pioneers during the development of the American chemical industry, was first clearly presented by A.D. Little in 1915:

Any chemical process, on whatever scale conducted, may be resolved into a coordinated series of what may be termed "unit actions," as pulverizing, mixing, heating, roasting, absorbing, condensing, lixiviating, precipitating, crystallizing, filtering, dissolving, electrolyzing, and so on. The complexity of chemical engineering results from the variety of conditions as to temperature, pressure, etc., under which the unit actions must be carried out in different processes and from the limitations as to materials of construction and design of apparatus imposed by the physical and chemical character of the reacting substances.

The original listing of the unit operations quoted above names twelve actions, not all of which are considered unit operations. Additional ones have been designated since then, at a modest rate over the years but recently at an accelerating rate.

1. Classification of unit operations

① Fluid flow. This concerns the principles that determine the flow or transportation of any fluid from one point to another.

② Heat transfer. This unit operation deals with the principles that govern accumulation and transfer of heat and energy from one place to another.

③ Evaporation. This is a special case of heat transfer, which deals with the evaporation of a volatile solvent such as water from a nonvolatile solute such as salt or any other material in solution.

④ Drying. In this operation volatile liquids, usually water, are removed from solid materials.

⑤ Distillation. This is an operation whereby components of a liquid mixture are separated by boiling because of their differences in vapor pressure.

⑥ Absorption. In this process a component is removed from a gas stream by treatment with a liquid.

⑦ Membrane separation. This process involves the diffusion of a solute from a liquid or gas through a semipermeable membrane barrier to another fluid.

⑧ Liquid-liquid extraction. In this case a solute in a liquid solution is removed by contacting with another liquid solvent which is relatively immiscible with the solution.

⑨ Liquid-solid leaching. This involves treating a finely divided solid with a liquid that dissolves out and removes a solute contained in the solid.

⑩ Crystallization. This concerns the removal of a solute such as a salt from a solution by precipitating the solute from the solution.

⑪ Mechanical physical separations. These involve separation of solids, liquids, or gases by mechanical means, such as filtration, settling, and size reduction, which are often classified as separate unit operations.

Many of these unit operations have certain fundamental and basic principles or mechanisms in common. For example, the mechanism of diffusion or mass transfer occurs in drying, absorption, distillation, and crystallization. Heat transfer occurs in drying, distillation, evaporation, and so on.

2. Fundamental concepts

Because the unit operations are a branch of engineering, they are based on both science and experience. Theory and practice must combine to yield designs for equipment that can be fabricated, assembled, operated, and maintained. The following four concepts are basic and form the foundation for the calculation of all operations.

(1) The material balance

If matter may be neither created nor destroyed, the total mass for all materials entering an operation equals the total mass for all materials leaving that operation, except for any material that may be retained or accumulated in the operation. By the application of this principle, the yields of a chemical reaction or engineering operation are computed.

In continuous operations, material is usually not accumulated in the operation, and a material balance consists simply of charging (or debiting) the operation with all material entering and crediting the operation with all material leaving, in the same manner as used by any accountant. The result must be a balance.

(2) The energy balance

Similarly, an energy balance may be made around any plant or unit operation to determine the energy required to carry on the operation or to maintain the desired operating conditions. The principle is just as important as that of the material balance, and it is used in the same way. The important point to keep in mind is that all energy of all kinds must be included, although it may be converted to a single equivalent form.

(3) The ideal contact

Whenever the materials being processed are in contact for any length of time under specified conditions, such as conditions of temperature, pressure, chemical composition, or electrical potential they tent to approach a definite condition of equilibrium which is determined by the specified conditions. In many cases the rate of approach to these equilibrium conditions is so rapid or the length of time is sufficient that the equilibrium conditions are practically attained at each contact. Such a contact is known as an equilibrium or ideal contact.

(4) Rate of an operation (the rate of transfer model)

In most operations equilibrium is not attained either because of insufficient time or because it is not desired. As soon as equilibrium is attained no further change can take place and the process stops, but the engineer must keep the process going. For this reason rate operations, such as rate of energy transfer, rate of mass transfer, and rate of chemical reaction, are of the greatest importance and interest. In all such cases the rate and direction depend upon a difference in potential or driving force. The rate usually may be expressed as proportional to a potential drop divided by a resistance. An application of this principle to electrical energy is the familiar Ohm' law for steady or direct current.

In solving rate problems as in heat transfer or mass transfer with this simple concept, the major difficulty is the evaluation of the resistance terms which are generally computed from an empirical correlation of many determinations of transfer rate under different conditions.

The basic concept that rate depends directly upon a potential drop and inversely upon a resistance may be applied to any rate operation, although the rate may be expressed in different ways with particular coefficients for particular cases.

(Geankoples Christie J. Transport Processes and Unit Operations. 2nd Edition.Allyn and Bacon Inc, 1983.)

Exercises

1. Put the following into Chinese

lixiviation	filter aid	flammability	isotope
sedimentation	settling	funnel	droplet
agglomeration	configuration	frag	vortex

2. Put the following into English

溶解	溶液	溶质	溶剂
提纯	不互溶的	浸取	过滤

Unit 2　Materials Science and Engineering

Lesson 1　Introduction to Materials Science and Engineering

Materials are properly more deep-seated in our culture than most of us realized. Transportation, housing, clothing, communication, recreation and food production – virtually every segment of our everyday lives is influenced to one degree or another by materials. Historically, the development and advancement of societies have been intimately tied to the members' abilities to produce and manipulate materials to fill their needs. In fact, early civilizations have been designated by the level of their materials development (i.e., Stone Age, Bronze Age).

The earliest humans have access to only a very limited number of materials, those that occur naturally, including stone, wood, clay, skins, and so on. With time they discovered techniques for producing materials that had properties superior to those of the natural ones: these new materials included pottery and various metals. Furthermore, it was discovered that the properties of a material could be altered by heat treatments and by the addition of other substances. At this point, materials utilization was totally a selection process, that is, deciding from a given, rather limited set of materials the one that was best suited for an application by virtue of its characteristic. It was not until relatively recent times that scientists came to understand the relationships between the structural elements of materials and their properties. This knowledge, acquired in the past 60 years or so, has empowered them to fashion, to a large degree, the characteristics of materials. Thus, tens of thousands of different materials have evolved with rather specialized characteristics that meet the needs of our modern complex society.

The development of many technologies that make our existence so comfortable has been intimately associated with the accessibility of suitable materials. Advancement in the understanding of a material type is often the forerunner to the stepwise progression of a technology.

Materials Science and Engineering

Material science is an interdisciplinary study that combines chemistry, physics, metallurgy, engineering and very recently life sciences. One aspect of materials science involves studying and designing materials to make them useful and reliable in the service of humankind.

It strives for basic understanding of how structures and processes on the atomic scale result in the properties and functions familiar at the engineering level. Materials scientists are interested in physical and chemical phenomena acting across large magnitudes of space and time scale. In this regard it differs from physicals or chemistry where the emphasis is more on explaining the properties of pure substances. In materials science there is also an emphasis on developing and using knowledge to understand how the properties of materials can be controllably designed by varying the compositions, structures, and the way in which the bulk and surfaces phase materials are processed.

Materials engineering is, on the basis of those structure properties correlations, designing or engineering the structure of a material to produce a predetermined set of properties. In the other words, materials engineering mainly deals with the use of materials in design and how materials are manufactured.

"Structure" is a nebulous term that deserves some explanation. In brief, the structure of a material usually relates to the arrangement of its internal components. Subatomic structure involves electrons within the individual atoms and interactions with their nuclei. On an atomic level, structure encompasses the organization of atoms or molecules relative to one another. The next large structural realm, which contains large groups of atoms that are normally agglomerated together, is termed "microscopic" meaning that which is subject to direct observation using some type of microscope. Finally, structural elements that may be viewed with the naked eye are termed "macroscopic".

The notion of "property" deserves elaboration. While in service use, all materials are exposed to external stimuli that evoke some type of response. For example, a specimen subject to forces will experience deformation; or a polished metal surface will reflect light. Property is a material trait in terms of the kind and magnitude of response to a specific imposed stimulus. Generally, definitions of properties are made independent of material shape and size.

Virtually all important properties of solid materials may be grouped into six different categories: mechanical, electrical, thermal, magnetic, optical, and deteriorative.

In addition to structure and properties, two other important components are involved in the science and engineering of materials namely "processing" and "performance". With regard to the relationships of these four components, the structure of a material will depend on how it is processed. Furthermore, a material's performance will be a function of its properties. Thus, the interrelationship between processing, structure, properties, and performance is linear as follows:

Processing→Structure→Properties→Performance

(William D Callister. Materials Science and Engineering: An Introduction, 2002.)

Exercises

1. Put the following into Chinese

| materials science | Stone age | naked eye | Bronze age |
| optical property | Integrated circuit | mechanical strength | thermal conductivity |

2. Put the following into English

| 交叉学科 | 介电常数 | 固体材料 | 热容 |
| 力学性质 | 电磁辐射 | 材料加工 | 弹性系数（模量） |

Lesson 2 Solid Materials and Engineering Materials

Solid materials are distinguished from the other states of matter (liquids and gases) by the fact that their constituent atoms are held together by strong interatomic forces. The electric and atomic structures, and almost all the physical properties, of solids depend on the nature and strength of primary interatomic bonds.

The technical materials used to build most structures are divided into three classes, metals, ceramics (including glasses), and polymers. These classes may be identified only roughly with the three types of interatomic bonding.

Metals: Materials that exhibit metallic bonding in the solid state are metals. Mixtures or solutions of different metals are alloys. About 85% of all metals have a crystal structure. In both face-centered cubic and hexagonal close-packed structures, every atom or ion is surrounded by twelve touching neighbors, which is the closest packing possible for spheres of uniform size. In any enclosure filled with close-packed spheres, 74% of the volume will be occupied by the spheres. In the body-centered cubic structure, each atom or ion has eight touching neighbors or eightfold coordination. Surprisingly, the density of packing is only reduced to 68% so that the BCC structure is nearly as densely packed as the FCC and HCP structure.

Ceramics: Ceramics materials are usually solid inorganic compounds with various combination of ionic or covalent bonding. They also have tightly packed structures, but with special requirement for bonding such as fourfold coordination for covalent solids and charge neutrality for ionic solids (i.e., each unit cell must be electrically neutral). As might be expected, these additional requirement lead to more open and complex crystal structures.

Carbon is often included with ceramics because of its much ceramic like properties, even though it is not a compound and conducts electrons in its graphitic form. Carbon is an interesting material since it occurs with two different crystal structures. In the diamond form, the four valence electrons of carbon lead to four nearest neighbors in tetrahedral coordination. This gives rise to the diamond cubic structure. An interesting variant on this structure occurs when the tetrahedral arrangement is distorted into a nearly flat sheet. The carbon atoms in the sheet have a hexagonal arrangement and stacking of the sheets gives rise to the graphite form of cardon. The (covalent) bonding within the sheets is much stronger than the bonding between sheets.

The existence of an element with two different crystal structures provides a striking opportunity to see how physical properties depend on atomic and electronic structure.

Inorganic glass: Some ceramic materials can be melted and upon cooling do not develop a crystal structure. The individual atoms have nearly the ideal number of nearest neighbors, but an orderly repeating arrangement is not maintained over long distances throughout the three-dimensional aggregates of atoms. Such noncrystals are called glasses or, more accurately, inorganic glasses and are said to be in the amorphous state. Silicates and phosphates, the two most common glass formers, have random three-dimensional network structures.

Polymers: The third category of solid materials includes all the polymers. The constituent atoms of classic polymers are usually carbon and are joined in a linear chairlike structure by covalent bonds .The bonds within the chain require two of the valence electrons of each atom, leaving the other two bonds available for adding a great variety of atoms (e. g., hydrogen), molecules, functional groups, and so on.

Based on the organization of these chains, there are two classes of polymers. In the first, the basic chains have little or no branching, such "straight" chain polymers can be melted and remelted without a basic change in structure (an advantage in fabrication) and are called thermoplastic polymers. If side chains are present and actually form (covalent) links between chains, a three-dimensional network structure is formed. Such structures are often strong, but once formed by heating will not melt uniformly on reheating. These are thermosetting polymers.

Usually both thermoplastic and thermosetting polymers have intertwined chains so that the resulting structures are quite random and are also said to be amorphous like glass, although only the thermosetting polymers have sufficient cross linking to form a three-dimensional network with covalent bonds. In amorphous thermoplastic polymers, many atoms in a chain are in close proximity to the atoms of adjacent chains, and van der Waals bonding and hydrogen bonding hold the chain together. They are these interchain bonds that are responsible for binding the substance together as a solid. Since these bonds are relatively weak, the resulting solid is relatively weak. Thermoplastic polymers generally have lower strengths and melting points than thermosetting polymers.

Engineering materials

In addition, there are three other groups of important engineering materials—

composites, semiconductors and biomaterials. Composites consist of combinations of two or more different materials, whereas semiconductors are utilized because of their unusual electrical characteristics; biomaterials are implanted into the human body.

① Composites: A number of composite materials have been engineered that consist of more than one material type. Fiberglass is a familiar example, in which glass fibers are embedded within a polymeric material. A composite is designed to display a combination of the best characteristics of each of the component materials. Fiberglass acquires strength from the glass and flexibility from the polymer. Many of the recent material developments have involved composite materials.

② Semiconductors: Semiconductors have electrical properties that are intermediate between the electrical conductors and insulators. Furthermore, the electrical characteristics of these materials are extremely sensitive to the presence of minute concentrations of impurity atoms, whose concentrations may be controlled over very small spatial regions. The semiconductors have made possible the advent of integrated circuitry that has totally revolutionized the electronics and computer industries over the past two decades.

③ Biomaterials: Biomaterials are employed in components implanted into the human body for replacement of diseased or damaged body parts. These materials most not produce toxic substances and must be compatible with body tissue (i.e., must not couse adverse biological reactions). All of the above materials—metals, ceramics, polymers, composites and semiconductors—may be used as biomaterials. For example, some of the biomaterials such as CF/C (carbon fiber/carbon) and CF/PS (polysulfone) are utilized in artificial hip replacements.

Advanced materials

Materials that are utilized in high technology (or high-tech) applications are sometimes termed advanced materials. By high technology we mean a device or product that operates or functions using relatively intricate and sophisticated principles; examples include electronic equipment (VCRs, CD players, etc.), computers, fiberoptic systems, spacecraft, air craft, and military rocketry. These advanced materials are typically either traditional materials whose properties have been enhanced or newly developed high-performance materials. Furthermore, they may be of all material types (e.g., metals, ceramics, polymers) and are normally relatively expensive.

Exercises

1. Put the following into Chinese

composite materials   advanced materials   transportation vehicles   nonrenewable resources organic compound　　　nuclear energy　　　raw materials　　　recycling technology

① Metals are extremely good conductors of electricity and heat, and are not transparent to visible light; a polished metal surface has a lustrous appearance.

② Ceramics are typically insulative to the passage of electricity and heat, and are more resistant to high temperatures and harsh environments than metals and polymers.

③ Materials processing and refinement methods need to be improved so that they produce less enviromental degradation, that is, less pollution and less despoilage of the landscape from mining of raw materials.

2. Put the following into English

工程材料　　　　可见光　　　　　非金属元素　　　　杂质
高科技　　　　　黏土矿物　　　　玻璃纤维　　　　　移植
材料加工　　　　环境质量

Lesson 3　An Introduction to Metallic Materials

What is a metal?

The key feature that distinguishes metals from non-metals is their bonding. Metallic materials have free electrons. In the case of pure metals, the outermost layer electrons are not bound to any given atom, instead these electrons are free to roam from atom to atom. Thus, the structure of metallic material can be thought of a consisting of positive centers (or irons) sitting in a "gas" of free-electrons.

The existence of this free electron gas has a number of profound consequences for the properties of metallic materials. For example, one of the most important features of metallic materials is that freely moving electrons can conduct electricity and so metallic materials tend to be good electrical conductors. Some metals more closely resemble the idealized picture of free electrons than others. Consequently, some metals are better conductors of electricity than others; for example, copper is a more efficient electrical conductor than tin. Electrical conductivity is such an important characteristic of metals that conductivity is sometimes used to distinguish metals from non-metals. The problem with using conductivity to distinguish metals from non-metals is that this approach is somewhat arbitrary. For example, graphite is a form of carbon which has quite a high electrical conductivity, but the bonding of carbon atoms in graphite is very different from that of atoms in metal. Therefore, it would be quite misleading to describe graphite as a metal. Note: the way in which atoms are arranged in a structure is just as important as the nature of the atoms themselves in determining the extent of electrical conductivity. Both diamond and graphite are made up of carbon, but diamond is a very good electrical insulator, rather than an electrical conductor like graphite.

What is an alloy?

An alloy consists of a mixture of a pure metal and one or more other elements. Often, these other elements will be metals. For example, brass is an alloy of copper and zinc. In other cases, a metal will be alloyed with a non-metal. The most important example of alloying involving addition of a non-metal would be (plain-carbon) steels, which consist of iron alloyed with carbon.

Alloys are usually less malleable and ductile than pure metals and they tend to have lower melting points. They do, however, have other properties which make them more useful than pure metals. An alloy is made by melting the different metals in the alloy together. The amounts of each metal are usually important.

Solid solutions and Intermetallic compounds

In many cases, metals are quite soluble in other metals. For example, solid copper and solid nickel are fully soluble in each other. This type of perfect solid solubility is a side effect of having free electrons. Since the electrons are free to move, the exact number of valence electrons possessed by any given atom shouldn't matter. Thus a metal should be able to dissolve another metal and produce a "solid-solution" in which one metal serves as the solvent and the other as the solute, although in a case like copper and nickel where these are mutually soluble at all compositions, the terms solvent and solute can be a little misleading.

In practice, however, not all metals are soluble in other metals. Thus, instead of a solid-solution a new phase, an "intermetallic compound", with a structure different from that of any of its constituent metals can be produced. For example, nickel will dissolve some aluminum, so that at low aluminum contents a solid solution is produced. However, if larger amounts of aluminum are added, then series of intermetallic compounds (for example Ni_3Al and $NiAl$) are produced. Some of these compounds (called "line compounds") have a very well defined composition (for example Ni_3Al invariably has almost exactly three nickel atoms for each aluminum atom). In contrast, other compounds have quite a wide range of composition (for example "NiAl" covers quite a wide range of nickel to aluminum ratios and there isn't necessarily exactly one nickel atom for every aluminum atom).

Exercises

1. Put the following into Chinese

① In the case of pure metals, the outermost layer electrons are not bound to any given atom, instead these electrons are free to roam from atom to atom.

② Alloys are usually less malleable and ductile than pure metals and they tend to have lower melting points.

③ The formation of intermetallic compounds seems strange, given the comment above about above free electrons promoting solid-solubility.

④ If two metals have different crystal structures then at some intermediate composition there will have to be a change from the crystal structure of one metal to that of the other.

2. Put the following into English

金属材料   溶质   溶剂   过渡金属
不锈钢   元素周期表   电负性   晶体结构

Lesson 4 Introduction to Ceramics

Definition

The word ceramic, derives its name from the Greek keramos, meaning "pottery", which in turn is derived from an older Sanskrit root, meaning "to burn". The Greek used the term to mean "burnt stuff" or "burned earth". Thus the word was used to refer to a product obtained through the action of fire upon earthy materials.

Ceramics make up one of three large classes of solid materials. The other material classes include metals and polymers. The combination of two or more of these materials together to produce a new material whose properties would not be attainable by conventional means is called a composite. Examples of composites include steel reinforced concrete, steel belted tyres, glass or carbon fibre-reinforced (so called fibre-glass resins) used for boats, tennis rackets, skis, and racing bikes.

Ceramics can be defined as inorganic, non-metallic materials that are typically produced using clays and other minerals from the earth or chemically processed powders. Ceramics are typically crystalline in nature and are compounds formed between metallic and non-metallic elements such as aluminium and oxygen (alumina, Al_2O_3), silicon and nitrogen (silicon nitride, Si_3N_4) and silicon and carbon (silicon carbide, SiC). Glass is somewhat different from ceramics in that it is amorphous, or has no long range crystalline order.

Most people, when they hear the word ceramics, think of art, dinnerware, pottery, tiles, brick and toilets. The above mentioned products are commonly referred to as traditional or silicate-based ceramics. While these traditional products have been, and continue to be, important to society, a new class of ceramics has emerged that most people are unaware of. These advanced or technical ceramics are being used for applications such as space shuttle tile, engine components, artificial bones and teeth, computers and other electronic components, and cutting tools, just to name a few.

Impact on society

We often take for granted the major role that ceramics have played in the progress of humankind. Let us look at a few examples of the importance of ceramics in our lives. Modern iron and steel and non-ferrous metal production would not be possible without the use of sophisticated refractory materials that are used to line high temperature furnaces, troughs and ladles. Metals make automobiles, machinery, planes, buildings and thousands of other useful things possible. Refractory ceramics are enabling materials for other industries as well. The chemical, petroleum, energy conversion, glass and other ceramic industries all rely on refractory materials.

Much of the construction industry depends on the use of ceramic materials. They include brick, cement, tile, and glass. Cement is used to make concrete which in turn is used for roadways, dams, buildings, and bridge. Uses of glass in the construction industry include various types of windows, glass block, and fibres for use in insulation, ceiling panels and roofing tiles. Brick is used for homes and commercial buildings because of its strength, durability, and beauty. Brick is the only building product that will not burn, melt, dent, peel, warp, rot, rust or be eaten by termites. Tile is used in applications such as flooring, walls, countertops, and fireplaces. Tile is also a very durable and hygienic construction product that adds beauty to any application.

An important invention that changed the lives of millions of people was the incandescent light bulb. This important invention by Thomas Edison in 1879 would not be possible without the use of glass. Glass's properties of hardness, transparency, and its ability to withstand high temperatures and hold a vacuum at the same time made the light bulb a reality. The evolution of lighting technology since this time has been characterized by the invention of increasingly brighter and more efficient light sources. By the middle of twentieth century, methods of lighting seemed well established—with filament and fluorescent lamps for interiors, neon lamps for exterior advertising and indicators, telecommunications (optical fibre networks), data storage (CD technology), and document production (laser printers).

The electronic industry would not exist without ceramics. Ceramics can be excellent insulators, semiconductors, superconductors, and magnets. It's hard to imagine not having mobile phones, computers, televisions, and other consumer electronic products. Ceramic spark plugs, while are electrical insulators, have had a large impact on society. They were first invented in 1986 to ignite fuel for internal combustion engines and are still being used for this purpose today. Applications include automobiles, boat engines, lawnmowers, and the like. High voltage insulators make it possible to safely carry electricity to houses and businesses.

The optical fibres have provided a technological breakthrough in the area of telecommunications. Information that was once carried electrically through hundreds of copper wires is now being carried through high-quality transparent silica (glass) fibres. Using this technology has increased the speed and volume of information that can be carried by orders of magnitude over that which is possible using copper cables. The reliability of the transmitted information is also greatly improved with optic fibres. In addition to these benefits, the negative effects of copper mining on the environment are reduced with the use of silica fibres.

Ceramics play an important role in addressing various environmental needs. Ceramics help decrease pollution, capture toxic materials and encapsulate nuclear waste. Today's catalytic converters in vehicles are made of cellular ceramics and help convert noxious hydrocarbons and carbon monoxide gases into non-toxic carbon dioxide and water. Advanced ceramic components are starting to be used in diesel and automotive engines. Ceramics' light weight and high-temperature and wear resistant properties, result in more efficient combustion and significant fuel savings.

Reusable, lightweight ceramic tile make NASA's space shuttle program possible. These thermal barrier tile protect the astronauts and the shuttle's aluminium frame from the extreme temperatures (up to approximately 1600℃) encountered upon re-entry into the earth's atmosphere.

(Greg Geiger. Journal of the American Ceramic Society, 1999.)

Words and Expressions

pottery　n. 陶器
refractory　adj., n. 难熔的，耐火材料
cement　n. 水泥，结合剂
incandescent　adj. 遇热发光的，白炽的
fluorescent　adj. 荧光的
diode　n. 二极管
encapsulate　vt. 装入胶囊，包裹
hygienic　adj. 卫生的
neon　n. 氖
noxious　adj. 有害的
trough　n. 出钢水口
ladle　n. 钢水包
termite　n. 白蚁
whisker　n. 晶须

agglomerate v. （使）聚集，团聚；n. 大块，聚集
suspension n. 悬浊液
dispersant n. 分散剂
heterogeneity n.不均匀性，多相性
granule n. 小（颗，细）粒
sanitary adj. （环境）卫生的，清洁的，保健的
viscous adj. 黏（性，稠）的
capillary n., adj. 毛细管（的）
abrasive n., adj. （研）磨料，摩擦力，磨料的
slump n., vi. 衰退，滑动，坍塌，落下
degradation n. 降低，下降，降解
sublimation n. 升华，提纯，纯化
slip casting 注浆成型
pressure casting 压力注浆成型
centrifugal casting 离心成型
injection molding 注模成型

Exercises

1. Reading comprehensions

① What kinds of materials are generally classified as ceramics?
② What are the properties of ceramic materials?

2. Translate the following into Chinese

amorphous materials   refractory materials   electrical insulator   silica fibre
thermal battier tile

3. Translate the following into English

碳纤维   非金属材料   类似黏土的材料   高炉温
混凝土

Reading material

Ceramic Processing Methods

Present methods of manufacturing ceramic green bodies of a complicated shape on an industrial level include dry-pressing with subsequent machining, slip casting, pressure casting, and injection molding. Tape casting is used to produce thin sheets, mainly for the electronic industry. All these forming methods start with a suspension where the ceramic particles (powders, whiskers, etc.) are mixed with a liquid or a polymer melt, proper dispersant, and possibly further additives such as binders, plasticizers, and antifoaming agents so that a well-dispersed, nonagglomerated ceramic slurry can be made.

Dry pressing and cold isostatic pressing (CIPing) are probably the most important forming techniques for industrial production of ceramic materials. Green bodies are formed by pressing granules in a die. The free-flowing granules are formed from a suspension using a granulation technique, e.g., spray drying or freeze granulating. Pressing is an established forming technique that has existed for decades and has been used for many applications, ranging from dinnerware to insulators and spark plugs. However there are developments in the field involving high-pressure CIPing and cyclic CIPing that can produce green bodies of higher density. The major advantage of dry pressing is productivity: modern presses can produce as many as 20 parts per minute. This makes pressing the method of choice for most industrial ceramic operations despite the problems associated with density gradients, inhomogeneous microstructure, and the need to machine complex-shaped objects.

All the drain-casting techniques e.g., slip casting, pressure casting, and centrifugal casting involve a solid-liquid separation process to form a dense green body. The liquid flow is driven by either an external pressure gradient (slip casting, pressuring casting) or a body force in a centrifugal force field (centrifugal casting). Slip casting is a low-pressure filtration method where capillary suction provides the driving force (on the order of 0.1-0.2MPa) for liquid removal and formation of a cast layer at a mold surface. Slip casting is generally a slow process, because the casting rate decreases parabolically with thickness of the cast layer. Pressure casting which is an established forming technique in fabrication of traditional clay-based ceramic materials, such as pottery and sanitary porcelain—and pressure filtration is modification of slip casting that has been developed to accelerate the consolidation stage and to obtain a higher green body density. In these methods, an external pressure (<4MPa) substantially higher than the capillary suction pressure is applied to the ceramic suspension.

The traditional drain-casting methods are plagued by some genetic problems. The liquid flow affects the suspension microstructure and tends to orient nonspherical constituents, such as whiskers. The stress gradient may also lead to nonuniform densities of the green body and cause mass segregation because of differences in particle size and density.

Undrained, or constant volume, forming methods, such as injection molding, have the potential to avoid the aforementioned problems. Injection molding is capable of producing parts of complex shape with high precision at relatively high production rates. This commonly used forming technique is based on mixing of the ceramics powder with a binder system (usually a mixture of polymers) to create a viscous feedstock and forming the part by injecting the powder/binder mixture into a impermeable mold, where the binder is solidified, usually by a temperature gradient. Injection molding has proved to be an excellent forming technique for smaller objects although there are potential problems related to the die-filling process.

The major problem confronting injection molding is the removal of the binder. Binder burnout must proceed at a slow rate (taking up to several days) to avoid problems with slumping and crack formation. The polymer removal time increases drastically when the size of the green body increases, making it difficult, if not impossible, to produce parts with thick cross sections. New systems with catalytic degradation of the polymer have been developed that have the potential to reduce many of the problems stated above through depolymerization and sublimation of the monomer at low temperatures. Using this catalytic degradation approach, the problems associated with the thermal expansion of the polymer, capillary forces, and particle migration due to liquid flow can be avoided. However, because of the high cost of the polymers used, this approach has found limited use.

The drying process has a major influence on green microstructure and production rate. Drying is a critical operation, which has to be controlled to avoid cracking and warping. Drying is a coupled heat- and mass-transfer problem for which mathematical representations have been available for years. The desired end in ceramic part production is fast drying; however, fast drying causes cracks. Tape-casting studies have shown that decreasing drying rates results in increasing green body densities, and binder additives strongly affect stress process. Cracking is inhibited by the solid net work, increasing pore size, and reducing capillary pressure. During drying, transport of evaporating dispersing media can cause binder and small particle migration to the surface. This can lead to additional problems during burnout and sintering. These problems can be minimized or avoided when the binder content is low or the dispersing media is sublimated.

(Sigmund W M, Bell N S, Bergstrom L. Journal of the American Ceramic Society, 2000.)

Lesson 5   Polymeric Composite Materials

Introduction

Principal advantage of composite materials resides in the possibility of combining physical properties of the constituents to obtain new structural or functional properties. Composite materials appeared very early in human technology, the "structural" properties of straw were combined with a clay matrix to produce the first construction material and, more recently, steel reinforcement opened the way to the ferroconcrete that is the last century dominant material in civil engineering. As a mater of fact, the modern development of polymeric materials and high modulus fibres (carbon, aramidic) introduced a new generation of composites. The most relevant benefit has been the possibility of energetically convenient manufacturing associated with the low weight features. Due to the possibility of designing properties, composite materials have been widely used, in the recent past, when stiffness/weight, strength/weight, ability to tailor structural performances and thermal expansion, corrosion resistance and fatigue resistance are required. Polymeric composites were mainly developed for aerospace applications where the reduction of the weight was the principal objective, irrespective of the cost.

Trends of composites

The main trends in the structural composite field are related to the reduction of the cost which cannot only be related to the improvement in the manufacturing technology, but needs an integration between design, material, process, tooling, quality assurance, manufacturing. Moreover, the high-tech industry, such as telecommunication, where specific functional properties are the principal requirements, will take advantages by the composite approach in the next future. The control of the filler size, shape and surface chemical nature has a fundamental role in the development of materials that can be utilized to develop devices, sensors and actuators based on the tailoring of functional properties such as optical, chemical and physical, magneto-elastic, etc. Finally, a future technological challenge will be the development of a new class of smart composite materials whose elasto-dynamic response can be adapted in real time in order to significantly enhance the performance of structural and mechanical systems under a diverse range of operating conditions.

Expectation and Needs for the next 10 years

The composite materials market is expected to expand in areas where costs are today a strong limitations. The improvement of the mature manufacturing technologies will benefit from integrated approach where product and process requirements are at design level. The expected reduction of manufacturing costs of the structural composites will expand applications of such materials to large-scale markets such as civil and goods.

Significant breakthroughs are expected in new composite materials especially in those applications, such as electronic, optic and biomedical, where functionality is the most relevant technical need. Relevant development will be expected in the area of nanophase material synthesis and nanocomposite manufacturing technology. However, further optimization studies are required to implement large-scale production.

Particular emphasis should be devoted to the R&D of composite materials able to respond to dynamic variation of the operative conditions. Smart materials will provide the nervous systems, the brains and the muscles for the existing advanced materials and structure that, at the moment, are a mere skeleton compared with the anatomy forecasted in a near future. Applications are expected in fields of sensors, actuators, and biomedical.

The quality of human life would be greatly improved by the availability of artificial prostheses and organs able to restore, repair or replace structural and functional performances of the natural tissues. The composite structure of the natural systems with its intrinsic complexity needs to be reproduced. Tissue engineering is one of the major focuses of biotechnological research today, with the expectation that this type of biohybrid technology will ultimately transform the practice of restorative clinics. The approach combines the principles of biology, material science and engineering to culture cells, also heterogeneous groups of cells, using polymeric biodegradable scaffolds as delivery vehicles for cell transplantation to obtain complex three-dimensional cellular constructs. Composite materials should be properly designed to provide anisotropic and/or active scaffolds able to control the cell growth in the reconstruction of complex natural structures.

The expected developments of the aforementioned fields need a serious interdisciplinary approach. As already recognized by US and Japan, significant advancements in the next 10 years in the field of (i) functional and structural composites through nanotechnologies, (ii) smart materials and (iii) composite for biomedical applications require cross-disciplinary strategies that should be addressed by combining various scientific disciplines. The available huge amount of human and economic resources, spread all over Europe, should be better coordinated by the creation of new interdisciplinary European research centers that could face the world scientific and technological challenge in these strategic fields.

Conclusions

Research activities, aimed to expand the applications in composite industry, must be addressed to improve manufacturing composite technology, through a better integration of product and process design; to develop new constituent materials with better performances and /or for the tailoring of structural and functional properties for special applications and for the development of new processes and new manufacturing technologies.

Expected breakthroughs are related to the development of multi-component materials with anisotropic and non-linear properties, able to impart unique structural and functional properties. Applications include smart systems, able to recognize and to adapt to external stimuli, as well as anisotropic and active composite systems to be used as scaffold for tissue engineering and other biomedical applications.

Exercises

1. Translate the following into Chinese

polymer composite   steel reinforcement   civil engineering   smart materials
corrosion resistance

2. Translate the following into English

环境友好的   生物可降解体系   纺织业   无线电通信
弹性动力学   重要进展   非线性的   耐疲劳强度

Lesson 6 Materials and Technology

Chemical research and development in the twentieth century have provided us with new materials that have profoundly improved the quality of our lives and helped to advance technology in countless ways. A few examples are polymers (including rubber and nylon), ceramics such as cookware, liquid crystals (like those in electronic displays), adhesives and coatings (for example, latex paint).

What is in store for the near future? One likely possibility is room-temperature superconductors. Electricity is carried by copper cables, which are not perfect conductors. Consequently, about 20 percent of electrical energy is lost in the form of heat between the power station and our homes. This is a tremendous waste. Superconductors are materials that have no electrical resistance and can therefore conduct electricity with no energy loss. Although the phenomenon of superconductivity at very low temperatures (more than 400 degrees Fahrenheit below the freezing point of water) has been known for over 80 years, a major breakthrough in the mid-1980s demonstrated that it is possible to make materials that act as superconductors at or near room temperature. Chemists have helped to design and synthesize new materials that show promise in this quest. The next 30 years high-temperature superconductors will be applied on a large scale in magnetic resonance imaging (MRI), levitated trains, and nuclear fusion.

If we had to name one technological advance that has shaped our lives more than any other, it would be the computer. The "engine" that drives the ongoing computer revolution is the microprocessor—the tiny silicon chip that has inspired countless inventions, such as laptop computers and fax machines. The performance of a microprocessor is judged by the speed with which it carries out mathematical operations, such as addition. The pace of progress is such that since their introduction, microprocessors have doubled in speed every 18 months. The quality of any microprocessor depends on the purity of the silicon chip and on the ability to add the desired amount of other substances, and chemists play an important role in the research and development of silicon chips. For the future, scientists have begun to explore the prospect of "molecular computing", that is, replacing silicon with molecules.

The advantages are that certain molecules can be made to respond to light, rather than to electrons, so that we would have optical computers rather than electronic computers. With proper genetic engineering, scientists can synthesize such molecules using microorganisms instead of large factories. Optical computers also would have much greater storage capacity than electronic computers.

Words and Expressions

ceramics   n. 制陶术，制陶业
adhesive   n. 黏合剂；adj, 带黏性的
latex   n. 乳汁，乳胶，橡胶
levitate   v. 使升空，使漂浮
magnetic resonance imaging   磁共振成像
microprocessor   n. 微处理器，单片机
laptop   n. 便携式电脑
budget   n. 预算；vi. 做预算
irrigation   n. 灌溉，冲洗
in addition   另外
keep in mind   谨记
break down   分解
not only…but also   不但……而且……

Lesson 7   Introduction to Nanoscale Materials

Introduction to the nanoworld

The nanoscale material with at least one dimension in the nanometer range is a bridge between isolated atoms or small molecules and bulk materials. Therefore, it is referred to as mesoscopic scale materials. Nanoscale materials as foundation of nanoscience and nanotechnology have become one of the most popular research topics in recent years. The intense interests in nanotechnology and nanoscale materials have paid to several areas by the tremendous economical, technological, and scientific impact: (i) with exponential growth of the capacity and speed of semiconducting chips, the key components which virtually enable all modern technology is rapidly approaching their limit of arts, this needs the coming out of new technology and new materials; (ii) novel nanoscale materials and devices hold great promise in energy, envirorunental, biomedical, and health sciences for more efficient use of energy sources, effective treatment of envirorunental hazards, rapid and accurate detection and diagnosis of human diseases; and (iii) when a material is reduced to the dimension of nanometer, its properties can be drastically different from those of the bulk material that we can either see or touch even though the composition is essentially the same. Therefore, nanoscale materials prove to be a very fertile ground for great scientific discoveries and explorations.

It has been said that a nanometer is "a magical point on the length scale", for this is the point where the smallest man-made devices meet the atoms and molecules of the natural world. Indeed, nanoscience and technology have been an explosive growth in the last few years.

Definition of nanoscale materials

(1) Nanometer

The prefix "nano" is from the Greek word "nanos" and it means dwarf. Nanometer is a length unit. A nanometer (nm) equals a billionth of a meter ($1nm = 1 \times 10^{-9}$ m).

Au atomic diameter is on the order of 0.1nm in size. The diameter of a carbon nanotube is about 1-2 nm, and a double helix of DNA is about 3 nm. A HIV virus is about 100 nm and so on. The diameter of one atom is about 0.1-0.2nm, and the length of 8-10 atoms is about one nanometer.

(2) nanoscale materials

Nanoscale material is defined as a material having one or more external dimensions in the nanoscale (1-100nm). A human hair is about 80000 nm in diameter. The single-walled carbon nanotube is about 0.1% of human hair in diameter.

Classification of nanoscale materials

Nanoscale materials are primitively divided into discrete nanomaterials and nanostructures materials, but also there are other classification methods.

The discrete nanomaterial means that the material has an appearance characteristic at least one dimension on the nanoscale, such as nanoparticles, nanofibers, nanotubes and membranes.

The nanostructures material is the material has an appearance characteristic of bulk material, but it may be built up of discrete nanomaterials, such as bulk materials by consolidation nanopowders, or it may be composed of continuously nanostructural units, such as porous materials including microporous (<2nm), mesoporous (2-50nm) and macroporous (>50nm), nanophase and polycrystalline materials.

The technique of consolidation nanopowders is a fabrication method of bulk nanostructures materials. However, because of the very small size of the powder particles, special precautions must be taken to reduce the interparticle frication and minimize the danger of explosion or fire. The powders themselves may have a microscale average particle size, or they may be true nanopowders, depending on their synthesis routes. They would be compacted at low or moderate temperature to produce a so-called green body with a density in excess of 90% of the theoretical maximum. Any residual porosity would be evenly distributed throughout the material and the pores would be fine in scale and have a narrow size distribution. Polycrystalline materials with grain sizes between 100 nm and 1μm are made up of many nanocrystals and are conventionally called ultrafine grains.

A reduction in the spatial dimension or confinement of nanoparticle in a particular crystallographic direction within a structure generally leads to changes in physical properties of the system in that direction. Hence one classification of nanostructures materials and systems essentially depends on the number of dimensions which lie within the nanometer range, as shown in Figure 4.

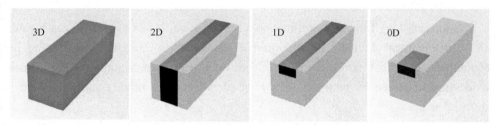

Figure 4   Dimensionality systems of the three-dimension (3D), two-dimension (2D), one-dimension (1D) and zero-dimension (0D)

(1) Zero-dimension (0D) materials

There are three dimensions for material on the nanoscale. This means that the size of material is confined in three dimensions (the material is dimensionless in three directions). This system includes the nanoparticles, nanocrystals, etc.

(2) One-dimension (1D) materials

There are two dimensions for material on the nanoscale. This means the size of material is confined in two dimensions (the material is dimensionless in the other two directions). The system includes nanowires, nanorods, nanofilaments, nanotubes, etc. The ratio of the length to the diameter of these structures is called aspect ratio. The aspect ratio for nanorods generally lies in the range of 10-100. If aspect ratio of nanorods becomes more than 100, they are termed as nanowires. Nanowires are hence long nanorods. Nanotubes are on the other hand, nanorods with hollow interiors.

(3) Two-dimension (2D) materials

There is one dimension on the nanoscale in material, that is, the size of material is confined in one dimension. The system includes ultrathin films, multilayered films, thin films, surface coatings, superlattices, etc.

Lesson 8 Applications of Nanomaterials

Nanotechnology encompasses a broad range of tools, techniques, and applications. This technology attempts to manipulate materials at the nanoscale in order to yield novel properties that do not exist at larger scales. These novel properties may enable new or improved solutions to problems that have been challenging to solve with conventional technology. For developing countries, these solutions may include more effective and inexpensive water purification devices, energy sources, medical diagnostic tests and drug delivery systems, durable building materials, and other products. It is quite apparent that there are innumerable potential benefits for society, the environment, and the world. Some of them have been briefly described as below.

Water purification

Nanotechnology for water purification has been identified as a priority area because the water treatment devices that incorporate nanoscale materials are already and the requirements of human development for clean water are pressing. Some water treatment devices manufactured by nanotechnology are already on the market and others are in advanced stages.

Nanofiltration membrane technology is already widely applied for removal of dissolved salts from salty or brackish water, removal of micro pollutants, water softening, and waste water treatment. Nanofiltration membranes are able to selectively remove harmful pollutants and retain nutrients present in water. It is expected that nanotechnology will contribute to improvements in membrane technology that will drive down the costs of desalination, which is currently a significant impediment to wider adoption of desalination technology.

Carbon nanotube filters offer a level of precision suitable for different applications as they can remove 25nm-sized polio viruses from water as well as larger pathogens such as E. coli and staphylococcus aureus bacteria. The nanotube-based water filters were found to filter bacteria and viruses, and were more resilient and reusable than conventional membrane filters. The filters were reusable by heating the nanotube filter or purging. Nano-engineered membranes allowed water to flow through the membrane faster than through conventional filters.

Nanocatalysts

Nanocatalysts include enzymes, metals, and other materials with enhanced catalytic capabilities that derive from either their nanoscale dimensions or from nanoscale structural modifications. Nanocatalysts such as titanium dioxide (TiO_2) and iron nanoparticles can be used to degrade organic pollutants and remove salts and heavy metals from liquids. People expect that nanoelectrocatalysts will enable the use of heavily polluted and heavily salinated water for drinking, sanitation, and irrigation.

Nanosensors

Nanosensors can detect single cell or even atom, making them far more sensitive than counterparts with larger components. Conventional water quality studies rely on a combination of on-site and laboratory analysis that requires trained staff to take water samples and access to a nearby laboratory to conduct chemical and biological analysis. New sensor technology combined with micro- and nanofabrication technologies is expected to lead to small, portable, and highly accurate sensors to detect chemical and biochemical parameters.

Medical applications

Nanotechnology offers a range of possibilities for health care and medicinal breakthroughs, including targeted drug delivery systems, extended-release vaccines and enhanced diagnostic and imaging technologies.

Nanoporous membranes may help with disease treatments in the developing world. They are a new way of slowly releasing a drug and are important for people far from hospitals. The nanopores are only slightly larger than that of molecules of drugs. Nanoporous membranes can control the rate of diffusion of the drug regardless of the amount of drug remaining inside a capsule. US nano-biopharmaceutical company NanoViricides Inc. has developed a viral therapy that uses an engineered flexible nanomaterial encapsulated active pharmaceutical ingredients to target specific viruses such as avian flu, common influenza and dismantle the viruses before they can infect cells.

Can it be that someday nanorobots will be able to travel through the body searching out and clearing up diseases, such as an arterial atheromatous plaque?

The greatest power of nanomedicine will emerge, perhaps in the 2020s, when we can design and construct complete artificial nanorobots using rigidly diamondoid nanometer-scale parts like molecular gears. These nanorobots will possess full panoply of autonomous subsystems including onboard sensors, motors, manipulators, power supplies, and molecular computers. But getting all these nanoscale components to spontaneously self-assemble in the right sequence will prove increasingly difficult as machine structures become more complex. Making complex nanorobotic systems requires manufacturing techniques that can build a molecular structure by what is called positional assembly. This will involve picking and placing molecular parts one by one, moving them along controlled trajectories much like the robot arms that manufacture cars on automobile assembly lines. The procedure is then repeated over and over with all the different parts until the final product, such as a medical nanorobot, is fully assembled.

Future nanorobots equipped with operating instruments and mobility will be able to perform precise and refined intracellular surgeries which are beyond the capabilities of direct manipulation by the human hand.

Unit 3 Polymer Science and Engineering

Lesson 1 Polymer

Polymers are all around us. They are the main components of food (starch, protein), clothes (silk, cotton, polyester, nylon), dwellings (wood-cellulose, paints) and also our bodies (nucleic acids, polysaccharides, proteins). No distinction is made between biopolymers and synthetic polymers. Indeed many of the early synthetic polymers were based upon naturally occurring polymers, e.g., celluloid (cellulose nitrate), vulcanization of rubber, rayon (cellulose acetate).

Polymers are constructed from monomer units, connected by covalent bonds. The definition of a polymer is:

"a substance, -R-R-R-R-, or in general -[R]$_n$-, where R is a bifunctional entity (or bivalent radical) which is not capable of a separate existence."

Where n is the degree of polymerization, DP$_n$. This definition excludes simple organic and inorganic compounds, e.g., CH_4, NaCl, and also excludes materials like diamond, silica and metals which appear to have the properties of polymer, but are capable of being vaporized into monomer units.

The molecular weight (MW) can be obtained from the MW of the monomer multiplied by n. When the value of n is small, say 2-20, the substances are called oligomers, often these oligomers are capable of further polymerizations and are then referred to as macromere.

A polymer with a MW of 10^7, if fully extended, should have a length of about 1mm and a diameter of about 0.5nm. This is equivalent in size to uncooked spaghetti about 2km in length. However, in reality, in bulk polymers the chain is never fully extended—a random coil configuration is adopted sweeping out a space of diameter about 200nm. It therefore has the appearance of cooked spaghetti or worms (or more correctly, worms of different length). The movements of these polymer chains are determined by several factors, such as:

① Temperature;

② Chemical make-up of the backbone —C—C—C— chain, whether the chain is flexible (aliphatic structure) or rigid (aromatic);

③ The presence or absence of side-chains on the backbone;

④ The inter-polymer chain attraction (weak-dipole/dipole, H-bonding, or strong covalent bonds, cross-linking);

⑤ The molecular weight (MW) and molecular weight distribution (MWD) of polymer.

Nearly all of the properties of polymers can be predicted if above factors are known, e.g., whether the polymer is amorphous or partially crystalline; the melting temperature of the crystalline phase (T_m) (actually it is more of a softening temperature over several degrees); whether the polymer is brittle or tough; its rigidity or stiffness (called modulus), whether the polymer dissolves in solvents, etc.

Polymers are really effect chemicals in that they are used as materials, e.g., plastics, fibers, films, elastomers, adhesives, paints, etc., with each application requiring different polymer properties. Many of the initial uses of plastics were inappropriate, which led to the belief that plastics were "cheap and nasty". However, recent legislation on product liability and a better understanding of the advantages and disadvantages of plastics have changed this position.

Economics that is the cost of making and fabricating the polymer is of prime importance. This has led to a rough grouping of polymers into commodity polymers, engineering polymers, and advanced polymeric materials.

1. Commodity polymers

Examples of these are:

i. Polyethylene $\begin{cases} \text{low density polyethylene (LDPE)} \\ \text{high density polyethylene (HDPE)} \\ \text{linear low density polyethylene (LLDPE)} \end{cases}$

ii. Polypropylene (PP)

iii. Poly vinyl chloride (PVC)

iv. Polystyrene (PS)

Each of these is prepared on the 10 million tones/year scale. The price is <$1500/tonne.

2. Engineering polymers

The materials have enjoyed the highest percentages growth of any polymers in the last ten years and are principally used as replacements for metals for moderate temperature (<150 ℃) and environmental conditions or they may have outstanding chemical inertness and/or special properties, e.g., low friction polytetrafluoroethylene (PTFE). These engineering polymers include:

i. Acetal (or polyoxymethylene, POM)

ii. Nylons (polyamides)

iii. Polyethylene or polybutylene terephthalate (PET or PBT)

iv. Polycarbonate (of bisphenol A) PC

v. Polyphenylene oxide (PPO)

The prices are ($3000-$15000)/tonne.

3. Advanced Polymeric Materials

These have very good temperature stability and when reinforced with fibers (e.g., glass, carbon or aramid fibres), i.e., composites, they are stronger than most metals on weight/weight basis. They are usually only used sparingly, often in critical parts of structure. The price can be as high as $150,000/tonne.

4. Making of Polymers

Approximately 100 million tones of polymers are made annually, in plants ranging from 240,000 tonnes/year continuous single stream polypropylene plants to a single batch preparation of a few kilograms of advanced performance composites. The highest tonnage polymers are LDPE, HDPE, LLDPE, PP, PVC, and PS.

The most important parameters in making polymers are quality control and reproducibility. They are different from simple organic compounds such as acetone, where often a simple distillation gives the desired purity. There are many different grades of the "same" polymers, depending on the final application, e.g., different MW, MWD, extent of branching, cross-linking, etc., and these variations are multiplied when copolymers (random, alternation and block) are considered. Many of these properties are fixed during polymerization and cannot be altered by post-treatment. Blending is sometimes carried out to obtain desired properties or just to up-grade production polymer that may be lightly off-specification.

A polymerization process consists of three stages:

① *Monomer preparation.* This is not discussed here, other than to emphasize that the purity of the monomer is paramount.

② *Polymerization.* As stated above, uniformity of polymer properties is absolutely necessary. The polymerization operation has to cope with the following parameters:

i. Homogeneous or heterogeneous reactions.

ii. In homogeneous system, control the viscosity.

iii. Most polymerizations are exothermic, heat removal should be conducted.

iv. Control of MW and MWD, branching and cross-linking…

③ *Polymer recovery.* Unless the polymerization takes place in bulk, separation from the solvent has to be carried out. The conventional methods of recovering chemicals, e.g., crystallization, distillation, adsorption, etc. are not be used because polymers possess properties such as high viscosity and low solubility in solvents, and are sticky and non-volatile. Nevertheless, precipitation by using a non-solvent followed by centrifuging, or by coagulation of an emulsion or latex and removal the solvent by steam-striping can be used.

5. Polymerization techniques

Most polymerizations are performed in the liquid phase using either a batch or a continuous process. The continuous method is preferred because it lends itself to smoother operation leading to a more uniform product, because of modern on-line analysis techniques. It also has lower operating cost. However, continuous processes have difficulties. The residence time of the polymer in the reactor will be variable, unless plug-flow is adopted using tube reactors. This may result in the following.

a. Catalyst residues in the polymer, e.g., peroxides, which may degrade the polymer during granulation or processing.

b. The polymer may have a broad MWD with some decomposition, e.g., cross-linking (for long residence times).

c. The polymer may adhere to the reactor walls, requiring shutdown for cleaning.

d. Repeated changes in polymer grade may be required, during change-over the polymer will be a mixture making the polymer unsuitable for use.

There are five general methods of polymerization:

a. Bulk (or mass).

b. Solution.

c. Slurry (or precipitation).

d. Suspension (or dispersion).

e. Emulsion.

Further lesser-used methods include:

a. Interfacial.

b. Reaction injection moulding (RIM).

c. Reactive processing of molten polymers.

(Brock William H. An Introduction to Industrial Chemistrym, The Fontana History of Chemistry. Fontana Press, 1992.)

Words and Expressions

dinitrogen   n. 分子氮，二氮
leguminous   adj. 豆科的，似豆科植物的
ambient   adj. 周围的，包围着的
bestow   vt. 把……赠予（给）
thumb one's noise (at)   （对……）做蔑视的手势
thermodynamics   n. 热力学
molybdenum   n. 钼 Mo
embed   vt. 把……嵌入，栽种

elude   vt. 使困惑，难倒
cobalt   n. 钴
hydrogenate   vt. 使与氢化合，使氢化
secondary reformer   二段（次）转化炉
ethanolamine   n. 乙醇胺
methanation   n. 甲烷化作用
plump   vi. 投票赞成，坚决拥护（for）
one-pass   单程，非循环过程
opt   vi. 选择，挑选（for, between）
acrylic   adj. 聚丙烯的，丙烯酸（衍生物）的
nitrile   n. 腈

Exercises

1. Put the following into Chinese

soda ash	refractory	silicate	chromatography
mercury	alkaline	desulphurisation	membrane
anode	cathode	contaminate	inert

2. Put the following into English

电解	分解	复分解	还原
沉淀	结晶	过滤	吸收
溶解度	溶度积	平衡	放热的

Lesson 2   What Are Polymers?

What are polymers? For one thing, they are complex and giant molecules and are different from low molecular weight compounds like, say, common salt. To contrast the difference, the molecular weight of common salt is only 58.5, while that of a polymer can be as high as several hundred thousands, even more than thousand thousands. These big molecules or "macro-molecules" are made up of much smaller molecules. The small molecules, which combine to form a big molecule, can be of one or more chemical compounds. To illustrate, imagine that a set of rings has the same size and is made of the same material. When these rings are interlinked, the chain formed can be considered as representing a polymer from molecules of the same compound. Alternatively, individual rings could be of different sizes and materials, and interlinked to represent a polymer from molecules of different compounds.

This interlinking of many units has given the polymer its name, poly meaning "many" and mer meaning "part" (in Greek). As an example, a gaseous compound called butadiene, with a molecular weight of 54, combines nearly 4000 times and gives a polymer known as polybutadiene (a synthetic rubber) with about 200000 molecular weight. The low molecular weight compounds from which the polymers form are known as monomers. The picture is simply as follows:

butadiene+ butadiene+…+butadiene⟶polybutadiene (4000 times)

One can thus see how a substance (monomer) with as small a molecular weight as 54 grows to become a giant molecule (polymer) of (54×4000≈)200000 molecular weight. It is essentially the "giantess" of the size of the polymer molecule that makes its behavior different from that of a commonly known chemical compound such as benzene. Solid benzene, for instance, melts to become liquid benzene at 5.5℃ and, on further heating, boils into gaseous benzene. As against this well-defined behavior of a simple chemical compound, a polymer like polyethylene does not melt sharply at one particular temperature into clean liquid. Instead, it becomes increasingly softer and, ultimately, turns into a very viscous, tacky molten mass. Further heating of this hot, viscous, molten polymer does convert it into various gases but it is no longer polyethylene (Figure 5).

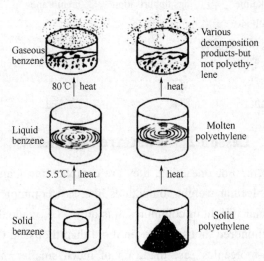

Figure 5 Difference in behavior on heating of a low molecular weight compound (benzene) and a polymer (polyethylene)

Another striking difference with respect to the behavior of a polymer and that of a low molecular weight compound concerns the dissolution process. Let us take, for example, sodium chloride and add it slowly to a fixed quantity of water. The salt, which represents a low molecular weight compound, dissolves in water up to a point (called saturation point) but, thereafter, any further quantity added does not go into solution but settles at the bottom and just remains there as solid. The viscosity of the saturated salt solution is not very much different from that of water. But if we take a polymer instead, say, polyvinyl alcohol, and add it to a fixed quantity of water, the polymer does not go into solution immediately. The globules of polyvinyl alcohol first absorb water, swell and get distorted in shape and after a long time go into solution. Also, we can add a very large quantity of the polymer to the same quantity of water without the saturation point ever being reached. As more and more quantity of polymer is added to water, the time taken for the dissolution of the polymer obviously increases and the mix ultimately assumes a soft, dough-like consistency. Another peculiarity is that, in water, polyvinyl alcohol never retains its original powdery nature as the excess sodium chloride does in a saturated salt solution. In conclusion, we can say that the long time taken by polyvinyl alcohol for dissolution, the absence of a saturation point, and the increase in the viscosity is all characteristics of a typical polymer being dissolved in a solvent and these characteristics are attributed mainly to the large molecular size of the polymer. The behavior of a low molecular weight compound and that of a polymer on dissolution are illustrated in Figure 6.

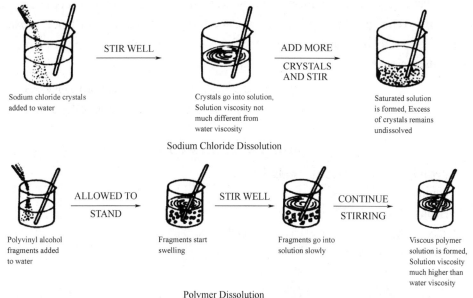

Figure 6 Difference in solubility behavior of a low molecular weight compound (sodium chloride) and a polymer (polyvinyl alcohol)

A polymer is a large molecule (macromolecule) composed of repeating structural units typically connected by covalent chemical bonds. While polymer in popular usage suggests plastic, the term actually refers to a large class of natural and synthetic materials with a variety of properties.

Due to the extraordinary range of properties accessible in polymeric materials, they have come to play an essential and ubiquitous role in everyday life—from plastics and elastomers on the one hand to natural biopolymers such as DNA and proteins that are essential for life on the other. A simple example is polyethylene, whose repeating unit is based on ethylene (IUPAC name ethene) monomer. Most commonly, as in this example, the continuously linked backbone of a polymer used for the preparation of plastics consists mainly of carbon atoms. However, other structures do exist; for example, elements such as silicon form familiar materials such as polysiloxanes, examples being silly putty and waterproof plumbing sealant. The backbone of DNA is in fact based on a phosphodiester bond, and repeating units of polysaccharides (e.g., cellulose) are joined together by glycosidic bonds via oxygen atoms.

Natural polymeric materials such as shellac, amber, and natural rubber have been in use for centuries. Biopolymers such as proteins and nucleic acids play crucial roles in biological processes. A variety of other natural polymers exist, such as cellulose, which is the main constituent of wood and paper.

The list of synthetic polymers includes synthetic rubber, bakelite, neoprene, nylon, PVC, polystyrene, polyethylene, polypropylene, polyacrylonitrile, PVB, polysiloxane, and many more. Polymers are studied in the fields of polymer chemistry, polymer physics, and polymer science.

(Gowariker V R, Viswanathan N V, Sreedhar J. Polymer Science. New York: John Wiley & Son, 1986.)

Words and Expressions

interlink   v. 把……互相连接起来；n. 连接
butadiene   n. 丁二烯
viscous   adj. 黏性的，黏的
tacky   adj. 俗气的，发黏的，缺乏教养或风度的
dissolution   n. 分解，溶解
saturation   n. 饱和
settle   vi. 解决，定居，沉淀
viscosity   n. [物] 黏性，[物] 黏度
globule   n. 水珠，药丸，血球，小球体
swell   vi. 膨胀，肿胀，隆起；vt. 使膨胀，使隆起；n. 肿胀，隆起
dough   n. 生面团
consistency   n. [计] 一致性，稠度，相容性
peculiarity   n. 特性，特质，怪癖，奇特
fragment   n. 碎片，片断或不完整部分

for one thing   首先
as against   和……比起来，和……相对照
convert…into…   把……转变成……
with respect to   关于
a quantity of …   大量，一些
be attributed to …   归因于，认为是……的结果

Exercises

1. Translate the following into Chinese

macromolecule	tacky	viscous	globule
settle	behavior	butadiene	powdery
molten polymer	distort	synthetic	fragment

2. Translate the following into English

氯化钠	单体	溶液	黏度
苯	形状	吸收	分子量
低分子化合物	溶胀	化合物	高分子化合物

Lesson 3   Polymer Synthesis

Polymers are a large class of materials consisting of many small molecules (called monomers) that can be linked together to form long chains, thus they are known as macromolecules. A typical polymer may include tens of thousands of monomers. Because of their large sizes, polymers are classified as macromolecules.

The study of polymer science begins with understanding the methods in which these materials are synthesized. Polymer synthesis is a complex procedure and can take place in a variety of ways.

Addition polymerization

Addition polymerization describes the method where monomers are added one by one to an active site on the growing chain. The most common type of addition polymerization is free radical polymerization. A free radical is simply a molecule with an unpaired electron. The tendency for this free radical to gain an additional electron in order to form a pair makes it highly reactive so that it breaks the bond on another molecule by stealing an electron, leaving that molecule with an unpaired electron (which is another free radical). Free radicals are often created by the division of a molecule (known as an initiator) into two fragments along a single bond. The following diagram shows the formation of a radical from its initiator, in this case benzoyl peroxide.

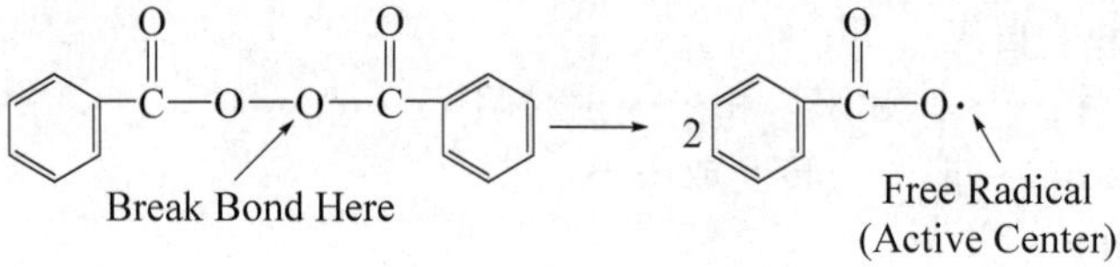

$$\text{Ph-C(=O)-O-O-C(=O)-Ph} \longrightarrow 2\ \text{Ph-C(=O)-O}\cdot$$

Break Bond Here → Free Radical (Active Center)

The stability of a radical refers to the molecule's tendency to react with other compounds. The stability of free radicals can vary widely depending on the properties of the molecule. An unstable radical will readily combine with many different molecules. However a stale radical will not easily interact with other chemical substances. The active center is the location of the unpaired electron on the radical because this is where the reaction takes place. In free radical polymerization, the radical attacks one monomer, and the electron migrates to another part of the molecule. This newly formed radical attacks another monomer and the process is repeated. Thus the active center moves down the chain as the polymerization occurs.

There are three significant reactions that take place in addition polymerization: initiation (birth), propagation (growth), and termination (death). These separate steps are explained below.

Initiation reaction

The first step in producing polymers by free radical polymerization is initiation. This step begins when an initiator decomposes into free radicals in the presence of monomers. The instability of carbon-carbon double bonds in the monomer makes them susceptible to reaction with the unpaired electrons in the radical. In this reaction, the active center of the radical "grabs" one of the electrons from the double bond of the monomer, leaving an unpaired electron to appear as a new active center at the end of the chain. Addition can occur at either end of the monomer.

In a typical synthesis, between 60% and 100% of the free radicals undergo an initiation reaction with a monomer. The remaining radicals may join with each other or with an impurity instead of with a monomer. "Self destruction" of free radicals is a major hindrance to the initiation reaction. By controlling the monomer to radical ratio, this problem can be reduced.

Propagation reaction

After a synthesis reaction has been initiated, the propagation reaction takes over. In the propagation stage, the process of electron transfer and consequent motion of the active center down the chain proceeds. In the following diagram, (chain) refers to a chain of connected monomers, and X refers to a substituent group (a molecular fragment) specific to the monomer. For example, if X was a methyl group, the monomer would be propylene and the polymer, polypropylene.

$$\text{\textasciitilde\textasciitilde}CH_2-\underset{X}{\underset{|}{C}}\overset{H}{\overset{|}{\cdot}} + \underset{H}{\overset{H}{>}}C=C\underset{X}{\overset{H}{<}} \longrightarrow \text{\textasciitilde\textasciitilde}H_2C-\underset{X}{\underset{|}{C}}\overset{H}{\overset{|}{-}}-CH_2-\underset{X}{\underset{|}{C}}\overset{H}{\overset{|}{-}}\text{\textasciitilde\textasciitilde}$$

In free radical polymerization, the entire propagation reaction usually takes place within a fraction of a second. Thousands of monomers are added to the chain within this time. The entire process stops when the termination reaction occurs.

Termination reaction

In theory, the propagation reaction could continue until the supply of monomers is exhausted. However this outcome is very unlikely. Most often the growth of a polymer chain is halted by the termination reaction. Termination typically occurs in two ways: combination and disproportionation.

Combination occurs when the polymer's growth is stopped by free electrons from two growing chains that join and form a single chain. The following diagram depicts combination, with the symbol (R) representing the rest of the chain.

$$R-H_2C-\underset{X}{\underset{|}{C}}\overset{H}{\overset{|}{\cdot}} + \overset{\cdot}{\underset{X}{\underset{|}{C}}}\overset{H}{\overset{|}{-}}CH_2-R \longrightarrow R-H_2C-\underset{X}{\underset{|}{C}}\overset{H}{\overset{|}{-}}-\underset{X}{\underset{|}{C}}\overset{H}{\overset{|}{-}}CH_2-R$$

Disproportionation halts the propagation reaction when a free radical strips a hydrogen atom form an active chain. A carbon-carbon double bond takes the place of the missing hydrogen. Termination by disproportionation is shown in the following diagram.

$$R-H_2C-\underset{X}{\underset{|}{C}}\overset{H}{\overset{|}{\cdot}} + \overset{\cdot}{\underset{X}{\underset{|}{C}}}\overset{H}{\overset{|}{-}}CH_2-R \longrightarrow R-H_2C-\underset{X}{\underset{|}{C}}\overset{H}{\overset{|}{-}}-H + C=CH-R$$

Disproportionation can also occur when the radical reacts with an impurity. This is why it is so important that polymerization be carried out under very clean conditions.

Polymerization

Polymerization is the process of combining many small molecules known as monomers into a covalently bonded chain. During the polymerization process, some chemical groups may be lost from each monomer. This is the case, for example, in the polymerization of PET polyester. The monomers are terephthalic acid ($HOOC-C_6H_4-COOH$) and ethylene glycol ($HO-CH_2-CH_2-OH$), but the repeating unit is $-OC-C_6H_4-$

COO-CH$_2$-CH$_2$-O-, which corresponds to the combination of the two monomers with the loss of two water molecules. The distinct piece of each monomer that is incorporated into the polymer is known as a repeat unit or monomer residue.

Laboratory synthesis

Laboratory synthetic methods are generally divided into two categories, step-growth polymerization and chain-growth polymerization. The essential difference between the two is that in chain growth polymerization, monomers are added to the chain one at a time only, whereas in step-growth polymerization chains of monomers may combine with one another directly. However, some newer methods such as plasma polymerization do not fit neatly into either category. Synthetic polymerization reactions may be carried out with or without a catalyst. Efferts towards rational synthesis of biopolymers via laboratory synthetic methods, especially artificial synthesis of proteins, is an area of intense research.

Biological synthesis

There are three main classes of biopolymers: polysaccharides, polypeptides, and polynucleotides. In living cells, they may be synthesized by enzyme-mediated processes, such as the formation of DNA catalyzed by DNA polymerase. The synthesis of proteins involves multiple enzyme-mediated processes to transcribe genetic information from the DNA to RNA and subsequently translate that information to synthesize the specified protein from amino acids. The protein may be modified further following translation in order to provide appropriate structure and function.

Modification of natural polymers

Many commercially important polymers are synthesized by chemical modification of naturally occurring polymers. Prominent examples include the reaction of nitric acid and cellulose to form nitrocellulose and the formation of vulcanized rubber by heating natural rubber in the presence of sulphur.

(Heaton C A. An Introduction to Industrial Chemistry. 2nd Edition. Blackie &Son Ltd, 1997.)

Words and Expressions

addition polymerization   n. 加成聚合反应
initiation   n. （链）引发
propagation   n. （链）增长
termination   n. （链）终止
combination   n. 耦合，组合，联合，化合
disproportionation   n. 歧化（作用或反应）

Exercises

Answer the following questions according to the text

① Give a brief explanation to the 3 main steps of addition polymerization.

② What is the difference between the two ways of termination: combination and disproportionation?

Lesson 4 Chain Polymerization and Bulk Polymerization

Chain polymerization

Many olefinic and vinyl unsaturated compounds are able to form chain-like macromolecules through elimination of the double bond, a phenomenon first recognized by Staudinger. Diolefins polymerize in the same manner, however, only one of the two double bonds is eliminated. Such reactions occur through the initial addition of a monomer molecule to an initiator or an initiator ion, by which the active state is transferred from the initiator to the added monomer. In the same way, by means of a chain reaction, one monomer molecule after the other is added until the active state is terminated through a different type of reaction. The polymerization is a chain reaction in two ways: because of the reaction kinetics and because as a reaction product one obtains a chain molecule. The length of the chain molecule is proportional to the kinetic chain length.

One thus obtains polyvinylchloride from vinylchloride, or polystyrene from styrene, or polyethylene, etc.

The length of the chain molecules, measured by means of the degree of polymerization, can be varied over a large range through selection of suitable reaction conditions. Usually, with commercially perpared and utilized polymers, the degree of polymerization lies in the range of 1 000 to 5 000, but in many cases it can be below 500 and over 10 000. This should not be interpreted to mean that all molecules of a certain polymeric material consist of 500, or 1 000, or 5 000 monomer units. In almost all cases, the polymeric material consists of a mixture of polymer molecules of different degrees of polymerization.

Polymerization, a chain reaction, occurs according to the same mechanism as wellknown chlorine-hydrogen reaction and the decomposition of phosgene.

The initiation reaction, which is the activation process pf the double bond, can be brought about by heating, irradiation, ultrasonics, or initiators. The initiation of the chain reaction can be observed most clearly with racial or ionic inititors. These are energy-rich compounds which can add suitable unsaturated compounds and maintain the activated radical, or ionic state so that further monomer molecules can be added in the same manner. For the individual steps of the growth reaction one needs only a relatively small activation enegy and therefore through a single activation step (the actual initiation reaction) a large number of olefin molecules are converted, as is implied by the term "chain reaction". Because very small amounts of the initiator bring about the formation of a large amount of polymeric (1:1000 to 1:10000), it is possible to regard polymerization from a superficial point of view as a catalytic reaction. For this reason, the initiators used in polymerization reactions are often designated as polymerization catalysts, even though, in the strictest sense, they are not true catalysts because the polymerization initiator enters into the reaction as a real partner and can be found chemically bound in the reaction product, i.e., the polymer.In addition to the ionic and radical initiators there are now metal complex initiators (which can be obtained, for example, by the reaction of titanium tetrachloride or titanium trichloride with aluminum alkyls), which play an important role in polymerization reactions (Ziegler catalysts). The mechanism of their catalytic action is not yet completely clear.

Bulk polymerization

Bulk polymerization traditionally has been defined as the formation of polymer from pure, undiluted monomers. Incidental amounts of solvents and small amounts of catalysts, promoters, and chain-transfer agents may also be present according to the classical definition. This definition however, serves little practical purpose. It includes a wide variety of polymers and polymerization schemes that have little in common, particularly from the viewpoint of reactor design. The modern gas-phase process for polyethylene satisfies the classical definition, yet is a far cry from the methyl methacrylate and styrene polymerization which remain single-phase throughout the polymerization and are more typically thought of as being bulk.

A common feature of most bulk polymerization and other processes not traditionally classified as such is the need to process fluids of very high viscosity. The high viscosity results from the presence of dissolved polymer in a continuous liquid phase. Significant concentrations of a high molecular-weight polymer typically increase fluid viscosities by 10^4 or more compared to the unreacted monomers. This suggests classifying a polymerization as bulk whenever a substantial concentration of polymer occurs in the continuous phase. Although this definition encompasses a wide variety of polymerization mechanisms, it leads to unifying concepts in reactor design. The design engineer must confront the polymer in its most intractable form, i.e., as a high viscosity solution or polymer melt.

The revised definition makes no sharp distinction between bulk and solution polymerizations and thus reflects industrial practice. Several so-called bulk processes for polystyrene and ABS use 5%-15% solvent as a processing aid and chain-transfer agent.

Few successful processes have used the very large amounts of solvent needed to avoid high viscosities in the continuous phase, although this approach is sometimes used for laboratory preparations.

Bulk polymerizations often exhibit a second, discontinuous phase. They frequently exhibit high exothermicity, but this is more characteristic of the reaction mechanism than of bulk polymerization as such. Bulk polymerizations of the free-radical variety are most common, although several commercially important condensation processes satisfy the revised definition of a bulk polymerization.

In all bulk polymerizations, highly viscous polymer solutions and melts are handled. This fact tends to govern the process design and to a lesser extent, the process economics. Suitably robust equipment has been developed for the various processing steps, including stirred-tank and tubular reactors, flash devolatilizers, extruder reactors, and extruder devolatilizers. Equipment costs are high based on working volume, but the volumetric efficiency of bulk polymerizations is also high. If a polymer can be made in bulk, manufacturing economics will most likely favor this approach.

It is tempting to suggest that polymer processes will gradually evolve toward bulk. Recently, the suspension process for impact polystyrene has been supplanted by the bulk process,and the emulsion process ABS may similarly be replaced. However, the modern gas-phase process for polyethylene appears to represent an opposite trend. It seems that polymerization technology tends to eliminate solvents and suspending fluids other than the monomers themselves. When the monomer is a solvent for the polymer, bulk processes as described in this article are chosen. When the monomer is not a solvent, suspension and slurry processes like those for polyethylene and polypropylene are employed. Hence, it is worthwhile avoiding a highly viscous continuous phase but not at the price of introducing extraneous material.

(Nauman E B.Enlyclopedia of polymer science and engineering. 2nd ed.Vol 2.Editor-in-chilf; Kroschwitz J I. New York :John Wiley & Sons, 1985:500～501)

Words and Expressions

chain polymerization　n. 链型聚合
bulk polymerization　n. 本体聚合

Lesson 5　Molecular Weight and Its Distributions of Polymers

The molecular weight of a polymer is of prime importance in its synthesis and application. The interesting and useful mechanical properties which are uniquely associated with polymeric materials are a consequence of their high molecular weights.

Most important mechanical properties depend on and vary considerably with molecular weight.

Thus, strength of polymer does not begin to develop until a minimum molecular weight of about 5000-10000 is achieved. Above that size, there is a rapid increase in the mechanical performance of polymers as their molecular weights increase; the effect levels off at still higher molecular weights. In most instance, there is some molecular weight range in which a given polymer property will be optimum for a particular application. The control of molecular weight is essential for the practical application of polymerization process.

When one speaks of the molecular weight of a polymer, one means something quite different from that which applies to small-sized compounds. Polymers differ from the small-sized compounds in that they are polydisperse or heterogeneous in molecular weight. Even if a polymer is synthesized free from contaminants and impurities, it is still not a pure substance in the usually accepted sense.

Polymers, in their purest form, are mixtures of molecules of different molecular weights. The reason for the polydispersity of polymers lies in the statistical variations present in the polymerization processes. When one discusses the molecular weight of a polymer, one is actually involved with its average molecular weight.

Both the average molecular weight and the exact distribution of different molecular weights within a polymer are required in order to fully characterize it. The control of molecular weight and molecular weight distribution (MWD) is often used to obtain and improve certain desired physical properties in a polymer product.

Various methods are available for the experimental measurement of the average molecular weight of a polymer sample. These include methods based on colligative properties, light scattering, viscosity, ultracentrifugation, and sedimentation. The various methods do not yield the same average molecular weight. Different average molecular weights are obtained because the properties being measured are biased differently toward the different sized polymer molecules in a polymer sample.

Some methods are biased toward the larger sized polymer molecules, while other methods are biased toward the smaller sized molecules. The result is that the average molecular weights obtained are correspondingly biased toward the larger or smaller sized molecules. The most important average molecular weights which are determined are the number-average molecular weight M_n, the weight-average molecular weight M_w and the viscosity-average molecular weight M_v.

In addition to the different average molecular weights of a polymer sample, it is frequently desirable and necessary to know the exact distribution of molecular weights. A variety of different fractionation methods are used to determine the molecular weight distribution of a polymer sample. These are based on fraction of a polymer sample using properties, such as solubility and permeability, which vary with molecular weight.

(Odian G. Principles of Polymerization. New York: McGraw-Hill Book Company, 1973:19)

Words and Expressions

mechanical property   n. 力学性能
strength   n. 强度
optimum   adj. 最适宜的；n. 最佳效果，最适宜条件
heterogeneous   adj. 多相的，不均匀的
contaminant   n. 污染物，致污物
statistical   adj. 统计的，统计学的
colligative   adj. 取决于分子的，依数的
ultracentrifugation   n. 超速离心法
sedimentation   n. 沉降，沉淀
viscosity average molecular weight   黏均分子量
fractionation   n. 分别，分馏法
solubility   n. 溶解度
permeability   n. 渗透性，透磁率
be biased toward (s)...   有……偏向，偏于……

Exercises

1. Translate the following into Chinese

polyvinyl alcohol   acetal   polyethylene   polyoxymethylene
epoxy resin   polyether   polyurethane   polyisobutylene
polyamide   polyvinylchloride   polybutadiene

2. Translate the following into English

离子型聚合   阴离子型聚合   阳离子型聚合   自由基聚合
加成聚合   逐步聚合   共聚合   均聚
缩聚   溶液聚合   乳液聚合   链式聚合

Lesson 6 Structure and Properties of Polymers

Most conveniently, polymers are generally subdivided in three categories, viz., plastics, rubbers and fibers. In terms of initial elastic modulus, rubbers ranging generally between 10^6 to 10^7 dynes/cm^2, represent the lower end of the scale, while fibers with high initial elastic modulus of 10^{10} to 10^{11} dynes/cm^2 are situated on the upper end of the scale; plastics, having generally an initial elastic modulus of 10^8 to 10^9 dynes/cm^2, lie in-between. As is found in all phases of polymer chemistry, there are many exceptions to this categorization.

An elastomer (or rubber) results from a polymer having relatively weak interchain forces and high molecular weight. When the molecular chains are "straightened out" or stretched by a process of extension, they do not have sufficient attraction for each other to maintain the oriented state and will retract once the force is released. This is the basis of elastic behavior.

However, if the interchain forces are very great, a polymer will make a good fiber. Therefore, when the polymer is highly strenched, the oriented chain will come under the influence of the powerful attractive forces and will "crystallize" permanently in a more or less oriented matrix. These crystallization forces will then act virtually as crosslinks, resulting in a material of high tensile strength and high initial molulus, i.e., a fiber. Therefore, a potential fiber polymer will not become a fiber unless subject to a "drawing" process, i.e., a process resulting in a high degree of intermolecular orientation.

Crosslinked species are found in all three categories and the process of crosslinking may change the cited characteristics of the categories. Thus, plastics are known to possess a marked range of deformability in the order of 100% to 200%; they do not exhibit this property when crosslinked, however. Rubber, on vulcanization, changes its properties from low modulus, low tensile strength, low hardness, and high elongation to high modulus, high tensile strength, high hardness, and low elongation. Thus, polymers may be classified as noncrosslinked and crosslinked, and this definition agrees generally with the subclassification in the thermoplastic and thermoset polymers. From the mechanistic point of view, however, polymers are properly divided into addition polymers and condensation polymers. Both of these species are found in rubbers, plastics, and fibers.

In many cases polymers are considered from the mechanistic point of view. Also, the polymer will be named according to its source whenever it is derived from a specific hypothetical monomer, or when it is derived from two or more components which are built randomly into the polymer.

This classification agrees well with the presently used general practice. When the repeating unit is composed of several monomeric components following each other in a regular fashion, the polymer is commonly named according to its structure.

It must be borne in mind that, with the advent of Ziegle-Natta mechanisms and new techniques to improve and extend crystallinity, and the closeness of packing of chains, many older data given should be critically considered in relation to the stereoregular and crystalline structure.

The properties of polymers are largely dependent on the type and extent of both stereoregularity and crystallinity.

Chemical properties

The attractive forces between polymer chains play a large part in determining a polymer's properties. Because polymer chains are so long, these interchain forces are amplified far beyond the attractions between conventional molecules. Different side groups on the polymer can lend the polymer to ionic bonding or hydrogen bonding between its own chains. These stronger forces typically result in higher tensile strength and higher crystalline melting points.

The intermolecular forces in polymers can be affected by dipoles in the monomer units. Polymers containing amide or carbonyl groups can form hydrogen bonds between adjacent chains; the partially positively charged hydrogen atoms in N—H groups of one chain are strongly attracted to the partially negatively charged oxygen atoms in C=O groups on another. These strong hydrogen bonds, for example, result in the high tensile strength and melting point of polymers containing urethane or urea linkages. Polyesters have dipole-dipole bonding between the oxygen atoms in C=O groups and the hydrogen atoms in H—C groups. Dipole bonding is not as strong as hydrogen bonding, so a polyester's melting point and strength are lower than Kevlar's (Twaron), but polyesters have greater flexibility.

Ethene, however, has no permanent dipole. The attractive forces between polyethylene chains arise from weak van der Waals forces. Molecules can be thought of as being surrounded by a cloud of negative electrons. As two polymer chains approach, their electron clouds repel one another. This has the effect of lowering the electron density on one side of a polymer chain, creating a slight positive dipole on this side. This charge is enough to attract the second polymer chain. van der Waals forces are quite weak, however, so polyethene can have a lower melting temperature compared to other polymers.

(Boening H V. Structure and Properties of Polymers. Geory Thieme Publishers Stuttgart, 1973: 18.)

Words and Expressions

elastic modulus   n. 弹性模量
stretch   vt. 伸展，张开；vi. 伸展；adj. 可伸缩的；n. 伸展，延伸
vulcanization   n. 橡胶的硫化
addition polymer   n. 加成聚合物
condensation polymer   n. 缩合聚合物
deformability   n. 可变形性
stereoregular   adj. 有规立构的

stereoregularity　n. 立构规整性
be derived from　来自于，由……派生而来
bear in mind　牢记，记住
with the advent of　随着……的出现
attraction for...　对……的引力
(be) subjected to　经受……，受到……

Exercises

1. Translate the following passage into Chinese

In general, head-to-tail addition is considered to be the predominant mode of propagation in all polymerizations. However, when the substituents on the monomer are small (and do not offer appreciable steric hindrance to the approaching radical) or do not have a large resonance stabilizing effect, as in the case of fluorine atoms, sizable amounts of head-to-head propagation may occur. The effect of increasing polymerization temperature is to increase the amount of head-to-head placement. Increased temperature leads to less selective (more random) propagation but the effect is not large. Thus the head-to-head content in poly (vinyl acetate) only increases from 1.30 to 1.98 percent when the polymerization temperature in increased from 30℃ to 90℃.

2. Translate the following into Chinese

entanglement	irregularity	sodium isopropylate	permeability
crystallite	stoichiometric balance	fractionation	light scattering
matrix	diffraction		

3. Translate the following into English

形态	酯化	杂质	二元胺
转化率	多分散性	构象	红外光谱法
高效液相法			

Lesson 7　Functional Polymers

Function polymers are macromolecules to which chemically functional groups are attached; they have the potential advantages of small molecules with the same functional groups. Their usefulness is related both to the functional groups and to the nature of the polymers whose characteristic properties depend mainly on the extraordinarily large size of the molecules.

The attachment of functional groups to polymer is frequently the first step toward the preparation of a functional polymer for a specific use. However, the proper choice of the polymer is an important factor for successful application. In addition to the synthetic aliphatic and aromatic polymers, a wide range of natural polymers have also been functionalized and used as reactive materials. Inorganic polymers have also been modified with reactive functional groups and used in processes requiring severe service conditions. In principle, the active groups and may be part of the polymer backbone or linked to a side chain as a pendant group either directly or via a spacer group. A required active functional group can be introduced onto a polymeric support chain ① by incorporation during the synthesis of the support itself through polymerization or copolymerization of monomers containing the desired functional groups, ② by chemical modification of a non-functionalized performed support matrix and ③ by a combination of ① and ②. Each of the two approaches has its own advantages and disadvantages, and one approach may be preferred for the preparation of a particular functional polymer when the other would be totally impractical. The choice between the two ways to the synthesis of functionalized polymers depends mainly on the required chemical and physical properties of the support for a specific application. Usually the requirements of the individual system must be thoroughly examined in order to take full advantage of each of the preparative techniques.

Rapid progress in the utilization of functionalized polymeric materials has been noted in the recent past. Interest in the field is being enhanced due to the possibility of creating systems that combine the unique properties of conventional active moieties and those of high molecular weight polymers. The successful utilizations of these polymers are based on the physical form, solution behavior, porosity, chemical reactivity and stability of the polymers. The various types of functionalized polymers cover a broad range of chemical applications, including the polymeric reactants, catalysts, carriers, surfactants, stabilizers, ion-exchange resins, etc. In a variety of biological and biomedical fields, such as the pharmaceutical, agriculture, food industry and the like, they have become indispensable materials, especially in controlled release formulation of drugs and agrochemicals. Besides, these polymers are extensively used as the antioxidants, flame retardants, corrosion inhibitors, flocculating agents, antistatic agents and the other technological applications. In addition, the functional polymers possess broad application prospects in the high technology area as conductive materials, photosensitizers, nuclear track detectors, liquid crystals, the working substances for storage and conversion of solar energy, etc.

(Akelah A, Moet A. Functionalized Polymers and Their Applications. London: Chapman and Hall, 1990.)

Words and Expressions

porosity   n. 有孔性，多孔性
pendant group   侧基

surfactant   n. 表面活性剂
ion exchange resin   n. 离子交换树脂
indispensable   adj. 不可缺少的，绝对必要的
controlled release   n. 控制释放
corrosion inhibitor   n. 缓蚀剂
flocculating agent   n. 絮凝剂
photosensitizer   n. 光敏剂

Lesson 8 Preparations of Amino Resins in Laboratory

Amino resins are reaction products of amino derivatives with aldehydes under acidic or basic conditions. The most important representatives of this class are the urea-formaldehyde (UF) and melamine-formaldehyde resins.

Chemicals. Urea, formalin (37%), ethanol, 2 mol/L NaOH solution, 0.1 mol/L NaOH solution, 1 mol/L standard NaOH solution, 1 mol/L standard HCl solution, glacial acetic acid, furfuryl alcohol, triethanolamine, wood flour, calcium phosphate, ammonium chloride, 0.5 mol/L H_2SO_4 solution, Na_2SO_3, 1% ethanoic thymolphthalein, indicator solution, melamine, glycerol and monomethylolurea.

Apparatus. Flasks and beakers, 500-ml three-neck resin kettle, heating mantle, mechanical stirrer, condenser, Dean-Stark trap, over, universal indicator paper, test tubes, 250-ml volumetric flask, ice bath, 10-ml pipette, burette, oil bath, screwcap and jar.

Preparation of a UF resin under acidic conditions. In order to demonstrate the rapidity of the reaction of urea with formaldehyde under acidic conditions, mix 5 g of urea with 6 ml of formalin in a test tube, and shake the tube until the urea has dissolved. Adjust the pH of the solution to 4 by the addition of 4 drop of 0.5 mol/L H_2SO_4, and observe the time required for precipitation to occur. Remove part of the precipitate and compare its solubility in water with the ample of monomethylolurea.

Preparation of a urea-formaldehyde adhesive. 60 grams (1 mole) of urea and 137 g (1.7 mole) of formalin (37%) are charged into a 500-ml reaction kettle equipped with a mechanical stirrer and a reflux condenser. The pH of the mixture is adjusted with 2 mol/L NaOH solution to between 7 and 8 as determined by universal indicator paper, and the mixture is refluxed for 2 hrs. At each subsequent 0.5-hr interval until the water has been removed, determine the free-formaldehyde content of the mixture by the procedure indicated below.

After the mixture has refluxed for 2 hrs, a Dean-Stark trap is introduced between the flask and the reflux condenser. About 40 ml of water is distilled into the trap and is discarded. The solution is acidified with 5 drops of glacial acetic acid; 44 g of furfuryl alcohol and 0.55 g of triethanolamine are introduced into the reaction mixture, and the solution is heated at 90℃ for 15 min.

The mixture is cooled to room temperature. A 15 g sample of resin is removed and is mixed with a hardener composed of 1 g of wood flour, 0.05 g of calcium phosphate, and 0.2 g of ammonium chloride. The mixture is set aside to harden at room temperature. The remaining resin to which the hardened has not been added is placed in a screwcap jar and is submitted to the laboratory instructor.

Determination of the free formaldehyde content. Prepare 250 ml of a 1mol/L Na_2SO_3 solution, and neutralize the solution, so it produces faint blue color with thymolphthalein indicator solution. Add a weighed 2g to 3g sample of resin into 100 ml of water in a 250-ml Erlenmeyer flask, and swirl the contents of the flask until they are dissolved completely. If the resin will not dissolve, ethanol may be added to aid the solution process. Cool the solution to 4 ℃ in an ice bath. Place 25 ml of the 1 mol/L Na_2SO_3 solution in a 100 ml-beaker, and pipet 10.00 ml of a standardized 1 mol/L HCl solution into the beaker. Cool the solution to 4 ℃. Add 10 to 15 drops of thymolphthalein indicator solution to the flask containing the sample, and adjust the color of the solution to a faint blue with 0.1 mol/L NaOH. Immediately, transfer the acid-sulfite solution to the flask containing the sample, completing the transfer with cold water. Titrate the solution to the thymolphthalein endpoint with standard 1 mol/L NaOH solution.

$$CH_2O + NaSO_3 + H_2O \longrightarrow CH_2OHSO_3Na^+ + NaOH$$

Determine the free formaldehyde content (%) from the quantity of HCl required to neutralize the resin solution.

Preparation of a melamine-formaldehyde resin. To a 500-ml reaction kettle equipped with a mechanical stirrer and a condenser is added 63 g (0.5 mole) of melamine and 122 g (1.5 mole) of formalin (37%). The mixture is refluxed for 40 min. The content (%) of free formaldehyde should be determined but at 10-min intervals. The procedure for the free-formaldehyde determination is given above.

After 20-min heating of the sample, a Dean-Stark trap is inserted between the flask and the condenser, and 10 ml of water is distilled off. The uncured sample is placed in a screwcap jar and is submitted along with the cured resin to the laboratory instructor.

(McCaffery E L. Laboratory Preparation for Macromolecular Chemistry.
New York: McGraw-Hill Book Company, 1970: 160.)

Lesson 9　Processing and Fabrication of Thermoplastics

Processing and fabrication describe the conversion of materials from stock form (bar, rod, tube, pellet, sheet, and so on) to a more or less complicated artefact. Polymer materials have proved especially amenable to variety of extrusion and moulding techniques. In particular the following principal processing and fabrication operations for thermoplastics now enable products and components of complex shape to be mass-producted on a very large scale.

The most important processing and fabrication techniques for thermoplastics exploit their generally low melting temperatures and shape the materials from the melt. Extrusion and injection moulding are the most widely used processes. The screw extruder accepts raw thermoplastics material in pellet form and carries it through the extruder barrel; the material is heated by contact with the heated barrel surface and also by the mechanical action of the screw, and melts. The melt is compressed by the taper of the screw and is ultimately extruded through the shaped die to form tube, sheet, rod or perhaps an extrusion of more complicated profile. The screw extruder works continuously and the extruded product is taken off as it emerges from the die for reeling or cutting into lengths. It is important that the melt viscosity should be sufficiently high to prevent collapse or uncontrolled deformation of the extrudate when it leaves the die, and there may be water or air sprays at the outlet for rapid cooling. High melt viscosities can be obtained by using materials of high molar mass. The rate of cooling of an extrusion may determine the degree of crystallinity in a crystalline polymer, and hence affect mechanical and other properties.

Injection moulding describes a process in which polymer melt is forced into a mould, where it cools until solid. The mould then separates into two halves to allow the product to be ejected; subsequently the parts of the mould are clamped together once more, a further quantity of melted material is injected and the cycle repeated. The injection end of the machine is most commonly an Archimedean screw (similar to that of a screw extruder) which can produce, once per cycle, a shot of molten polymer of predetermined size and then inject it into the mould by means of a reciprocating ram action. Injection moulding provides a particularly effective way of obtaining complex shapes in large production runs.

As the size and/ or aspect ratio of an injection moulding increase it becomes more difficult to ensure uniformity in the polymer during injection and to maintain a sufficient clamping force to keep the mould closed during filling. The reaction injection moulding process has been developed to overcome these problems, essentially by carrying out most of the polymerizing reaction in the mould.

Blow moulding represents a development of extrusion in which hollow articles are fabricated by trapping a length of extruded tude (the parison) and inflating it within a mould. The simple extruded parison may be replaced by an injection-moulded preform.

Hollow articles including those of large dimensions may also be produced by rotational moulding (or rotocasting). A charge of solid polymer, usually powder, is introduced into a mould which is first heated to form a melt. The mould is then rotated about two axes to coat its interior surface to a uniform thickness.

Polymeric materials in continuous sheet form are often produced and subsequently reduced in thickness by passing between a series of heated rollers, an operation known as calendering. Thermoforming employs suction or air pressure to shape a thermoplastic sheet heated above its softening temperature to the contours of a male or female mould. Certain thermoplastics can be shaped without heating in a number of cold-forming operations such as stamping and forging commonly applied to metals.

All these methods of fabrication are essentially moulding processes. Cutting techniques (embracing all the conventional machining operations: turning, drilling, grinding, milling and planing) can also be applied to thermoplastics, but they are used much less widely. These techniques are most commonly applied to glassy polymers such as PMMA and to thermoplastic composites such as glass-filled PTFE, although softer materials such as PE and unfilled PTFE have excellent machinability.

Many thermoplastics may be satisfactorily cemented either to similar or dissimilar materials. A number of inert and insoluble polymer materials such as PTFE , PCTFE and some other fluoropolymers, PE, PP and some other polyolefins are amenable to cementing only after vigorous surface treatment .For the lower melting thermoplastics welding provides an important alternative for joining parts of the same material, for example in fabricating pipework. The thermoplastics of higher melting temperature (which include PTFE of the common engineering polymers and also some of the specialised heat-tolerant materials such as the polyimides) cannot be satisfactorily fabricated by the principal extrusion and injection moulding processes.　These materials are shaped by sintering powdered polymer in pressurized moulds, a process which causes the polymer particles to coalesce.

(Christopher Hall. Polymer Materials. London : The Macmillan Press Ltd , 1981:162-165.)

Words and Expressions

amenable　adj. 有责任的，应服从的，有义务的
thermoplastics　n. 热塑性塑料（thermoplastic 的复数）
screw extruder　n. 螺旋挤出机
moulding processes　n. 注模过程
injection moulding　n. 注塑
blow moulding　n. 吹塑

extrusion   n. 挤出，推出，赶出，喷出

Exercises

1. Translate the following into Chinese

collapse	rotocasting	heat-tolerant	clamp
sintering	reciprocating	welding	coalesce

2. Translate the following into English

螺杆挤出机	长径比	型坯	口模
反应注塑成型	压延	吹塑	粘接
挤出成型			

Unit 4   Pharmaceuticals Engineering

Lesson 1   Production of Drugs

Depending on their production or origin pharmaceutical agents can be split into three groups:

Ⅰ. Totally synthetic materials (synthetics);

Ⅱ. Natural products;

Ⅲ. Products from partial syntheses (semi-synthetic products).

The emphasis of the present book is on the most important compound of group Ⅰ and Ⅲ—thus Drug *synthesis*. This does not mean, however, that natural products or other agents are less important. They can serve as valuable lead structures, and they are frequently needed as starting materials or as intermediates for important synthetic products.

Table 4 gives an overview of the different methods for obtaining pharmaceutical agents.

Table 4   Possibilities for the preparation of drugs

Methods	Examples
1. Total synthesis	—over 75% of all pharmaceutical agents (synthetics)
2. Isolation from natural sources(natural products):	
2.1　Plants	—alkaloids; enzymes; heart glycosides; polysaccharides; tocopherol; steroid precursors (diosgenin, sitosterin); citral (intermediate product for vitamins A, E, and K)
2.2　Animal organs	—enzymes; peptide hormones ;cholic acid from gall; insulin from the pancreas; sera and vaccines
2.3　Other sources	—cholesterol from wool oils; L-amino acids from keratin and gelatine hydrolysates
3. Fermentation	—antibiotics; L-amino acids; dextran; targeted modifications on steroids, e.g. 11-hydroxylation; also insulin, interferon, antibodies, peptide hormones, enzymes, vaccines
4. Partial synthetic modification of natural products (semisynthetic agents):	
	—alkaloid compounds; semisynthetic β-lactam antibiotics; steroids; human insulin

Several therapeutically significant natural products which were originally obtained from natural sources are today more effectively – i.e., more economically – prepared by total synthesis. Such examples include L-amino acids, Chloramphenicol, Caffeiene, Dopamine, Epinephrine, Levodopa, Peptide hormones, Prostaglandins, D-Penicillamine, Vincamine, and practically all vitamins.

Over the last few years fermentation – i.e., microbiological processes – has become extremely important. Through modern technology and results from genetic selection leading to the creation of high performance mutants of microorganisms, fermentation has already become the method of choice for a wide range of substances. Both Eukaryotes (yeasts and moulds) and Prokaryotes (single bacterial cells, actinomycetes) are used as microorganism. The following product types can be obtained:

Ⅰ. Cell material;

Ⅱ. Enzymes;

Ⅲ. Primary degradation products;

Ⅳ. Secondary degradation products.

Disregarding the production of dextran from the mucous membranes of certain microorganisms, e.g., *Leuconostoc mesenteroides*, classes Ⅱ and Ⅲ are the relevant ones for the preparation of drugs. Dextran itself, with a molecular weight of 50,000-100000, is used as a blood plasma substitute. Among the primary metabolites the L-amino acids from mutants of *Corynebacterium glutamicum* and *Brevibacterium flavum* are especially interesting. From these organisms some 350000 tones of monosodium L-glutamate (food additive) and some 70000 tones of L-lysine (supplement for vegetable proteins) are produced. Further important primary metabolites are the Purine nucleotides, organic acids, lactic acid, citric acid, and vitamins, for example vitamin B_{12} from *Propionibacterium shermanii*.

Among the secondary metabolites the antibiotics must be mentioned first. The following five groups represent a yearly worldwide value of US-$ 17 billion:penicillins (*Penicillium chrysogenum*), cephalosporins (*Cephalosporium acremonium*), tetracyclines (*Streptomyces aureofaciens*), erythromycins (*Streptomyces erythreus*), aminoglycoseds (e.g., streptomycin from *Streptomyces griseus*).

About 5000 antibiotics have already been isolated from microorganisms, but of these only somewhat fewer than 100 are in therapeutic use. It must be remembered, however, that many derivatives have been modified by partial synthesis for therapeutic use; some 50000 agents have been semisynthetically obtained from β-lactams alone in the last decade. Fermentations are carried out in stainless steel fermentors with volumes up to $400m^3$. To avoid contamination of the microorganisms with phages, etc., the whole process has to be performed under sterile conditions. Since the more important fermentations occur exclusively under aerobic conditions a good supply of oxygen or air (sterile) is needed. Carbon dioxide sources include carbohydrates, e.g., molasses, saccharides, and glucose. Additionally the microorganisms must be supplied in the growth medium with nitrogen-containing compounds such as ammonium sulfate, ammonia, or urea, as well as with inorganic phosphates. Furthermore, constant optimal pH and temperature are required. In the cases of penicillin G, the fermentation is finished after 200 hours, and the cell mass is separated by filtration. The desired active agents are isolated from the filtrate by absorption or extraction processes. The cell mass, if not the desired product, can be further used as an animal feedstuff owing to its high protein content.

By modern recombinant techniques microorganisms have been obtained which also allow production of peptides which were not encoded in the original genes. Modified E. coli bacteria make it thus possible to produce A- and B- chains of human insulin or proinsulin analogs. The disulfide bridges are formed selectively after isolation, and the final purification is effected by chromatographic procedures. In this way human insulin is obtained totally independently from any pancreatic material taken from animals.

Other important peptides, hormones, and enzymes, such as human growth hormone (HGH), neuroactive peptides, somatostatin, interferons, tissue plasminogen activator (TPA), lymphokines, calcium regulators like calmodulin, protein vaccines, as well as monoclonal antibodies used as diagnostics, are synthesized in this way.

The enzymes or enzymatic systems which are present in a single microorganism can be used for directed stereospecific and regiospecific chemical reactions. This principle is especially useful in steroid chemistry. Here we may refer only to the microbiological 11-α-hrdro xylation of progesterone to 11-α-hydroxyprogesterone, a key product used in the synthsis of cortisone. Isolated enzymes are important today not only because of the technical importance of the enzymatic saccharification of starch, and the isomerization of glucose to fructose. They are also significant in the countless test procedures used in diagnosing illness, and in enzymatic analysis which is used in the monitoring of therapy.

A number of enzymes are themselves used as active ingredients. Thus preparations containing proteases (e.g., chymotrypsin, pepsin and trypsin), amylases and lipases, mostly in combination with synthetic antacids, promote digestion. Streptokinase and urokinase are important in thrombolytics, and asparaginases is used as a cytostatic agent in the treatment of leukemia.

Finally mention must be made of the important use of enzymes as "biocatalysts" in chemical reactions where their stereospecificity and selectivity can be used. Known examples are the enzymatic cleavage of racemates of N-acetyl-D, L-amino acid to give L-amino acid, the production of 6-aminopenicillanic acid from benzylpenicillin by means of penicillinamidase and the aspartase-catalysed stereospecific addition of ammonia to fumaric acid in order to produce L-aspartic acid.

In these applications the enzymes can be used in immobilized forms — somehow bound to carriers—and so used as heterogeneous catalysts. This is advantageous because they can then easily be separated from the reaction medium and recycled for further use.

Another important process depending on the specific action of proteases is applied for the production of semisynthetic human insulin. This starts with pig insulin in which the alanine in the 30-position of the B-chain is replaced by a threonine tert-butyl ester by the selective action of trypsin. The insulin ester is separated, hydrolyzed to human insulin and finally purified by chromatographic procedures.

Sources for enzymes include not only microorganisms but also vegetable and animal materials.

In table 4 it was already shown that over 75% of all pharmaceutical agents are obtained by total synthesis. Therefore knowledge of the synthetic routes is useful. Understanding also makes it possible to recognize contamination of the agents by intermediates and by-products. For the reason of effective quality control the registration authorities in many countries demand as essentials for registration a thorough documentation on the production process. Knowledge of drug syntheses provides the R&D chemist with valuable stimulation as well.

There are neither preferred structure classes for all pharmaceutically active compounds nor preferred reaction types. This implies that practically the whole organic field and in part also organometallic chemistry is covered. Nevertheless, a larger number of starting materials and intermediates are more frequently used, and so it is useful to know the possibilities for their preparation from primary chemicals. For this reason it is appropriate somewhere in this book to illustrate a tree of especially important intermediates. These latter intermediates are the key compounds used in synthetic processes leading to an enormous number of agents. For the most part chemicals are involved which are produced in large amounts. In a similar way this is also true for the intermediates based on the industrial aromatic compounds toluene, phenol and chlorobenzene. Further key compounds may be shown in a table which can be useful in tracing cross-relationships in syntheses.

In addition to the actual staring materials and intermediates solvents are required both as a reaction medium and for purification via recrystallization. Frequently used solvents are methanol, ethanol, isopropanol, butanol, acetone, ethyl acetate, benzene, toluene and xylene. To a lesser extent diethyl ether, tetrahydrofuran, glycol ethers, dimethylformamide (DMF) and dimethyl sulphoxide (DMSO) are used in special reactions.

Reagents used in larger amounts are not only acids (hydrochloric acid, sulfuric acid, nitric acid, acetic acid) but also inorganic and organic bases (sodium hydroxide, potassium hydroxide, potassium carbonate, sodium bicarbonate, ammonia, triethylamine, pyridine). Further auxiliary chemicals include active charcoal and catalysts. All of these supplementary chemicals (like the intermediates) can be a source of impurities in the final product.

(Roth H J, Kleemann A. Pharmaceutical Chemistry. Vol 1, Drug Synthesis, England: Ellis Horwood Limited, 1988.)

Words and Expressions

pharmaceutical   adj. 制药的，药学的；n. 药品，药剂
alkaloid   n. 生物碱
enzyme   n. 酶
polysaccharide   n. 多糖，多聚糖
precursor   n. 前体
steroid   n. 甾体
peptide   n. 肽，缩氨酸
hormone   n. 激素，荷尔蒙
gall   n. 胆汁
insulin   n. 胰岛素
pancreas   n. 胰腺
serum   n. 血浆
vaccine   n. 疫苗，牛痘疫苗
cholesterol   n. 胆固醇
gelatine   n. 骨胶，明胶
antibiotic   adj. 抗生的，抗菌的；n. 抗生素
interferon   n. 干扰素
antibody   n. 抗体
fermentation   n. 发酵
therapeutical   adj. 治疗（学）的
caffeine   n. 咖啡因，咖啡碱
dopamine   n. 多巴胺（一种神经递质）
yeast   n. 酵母
mucous   adj. 黏液的，分泌黏液的
plasma   n. 血浆，淋巴液，等离子体
penicillin   n. 青霉素

streptomycin   n. 链霉素
derivative   n. 衍生物
sterile   adj. 不能生育的，无细菌的
aerobic   adj. 需氧的，有氧的
feedstuff   n. 饲料
lymph   n. 淋巴，淋巴液
starch   n. 淀粉
regiospecific reaction   区域专一性反应
stereospecific reaction   立体专一性反应
glucose   n. 葡萄糖
streptokinase   n. 链球葡萄激酶
immobilize   vt. 固定化
heterogeneous   adj. 不均匀的，多相的
trypsin   n. 胰蛋白酶
contamination   n. 沾污，污染，污染物
hygienic   adj. 卫生学的，卫生的
intemediate   n. 中间体
extraction   n. 萃取，提取
recrystallization   n. 重结晶
xylene   n. 二甲苯
toluene   n. 甲苯
ether   n. 醚
benzene   n. 苯

Exercises

1. Answer the following questions

① How many groups can pharmaceutical agents be split into depending on their production or origin?

② Can you illustrate any significant examples of pharmaceutical agents obtained by total synthesis?

③ What is the difference between the synthetic drugs and traditional Chinese herbal medicine?

2. Put the following into English

生物碱	中间体	起始原料	重结晶
胆固醇	吡啶	甲苯	萃取
胰岛素	醛		

3. Put the following into Chinese

polysaccharide	peptide	hormone	vaccine
heterogeneous catalyst	contamination	plasma	steroid
penicillin	metabolite		

Lesson 2 Isolation of Caffeine from Tea

In this experiment, caffeine will be isolated from tea leaves. The major problem of the isolation is that caffeine does not occur alone in tea leaves, but is accompanied by other natural substances from which it must be separated. The major component of tea leaves is cellulose, which is the major structural material of all plant cells. Cellulose is a polymer of glucose. Since cellulose is virtually insoluble in water, it presents no problems in the isolation procedure. Caffeine, on the other hand, is water soluble and is one of the major substances extracted into the solution called "tea". Caffeine comprises as much as 5% by weight of the leaf material in tea plants. Tannins also dissolve in the hot water used to extract tea leaves. The term tannins does not refer to a single homogeneous compound, or even to substances which have similar chemical structure. It refers to a class of compounds which have certain properties in common. Tannins are phenolic compounds having molecular weights between 500 and 3000. They are widely used to "tan" leather. They precipitate alkaloids and proteins from aqueous solutions. Tannins are usually divided into two classes: those which can be hydrolyzed and those which cannot. Tannins of the first type which are found in tea generally yield glucose and gallic acid when they are hydrolyzed. These tannins are esters of gallic acid and glucose. They represent structures in which some of the hydrolyzable groups in glucose have been esterified by digalloyl groups. The non-hydrolyzable tannins found in tea are condensation polymers of catechin. These polymers are not uniform in structure, but catechin molecules are usually linked together at ring position 4 and 8. The following is their structures.

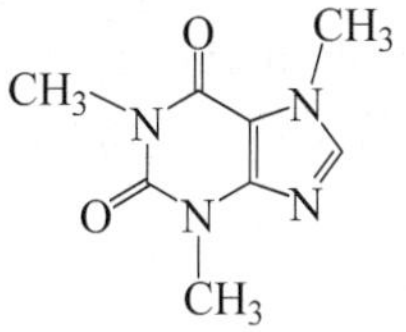

Caffeine

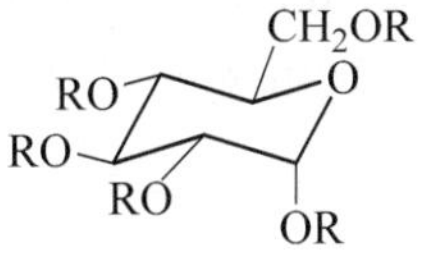

Glucose if R=H
A Tannin if some R=Digalloyl

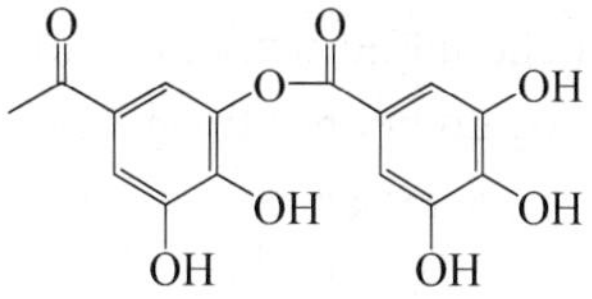

A Digalloyl Group

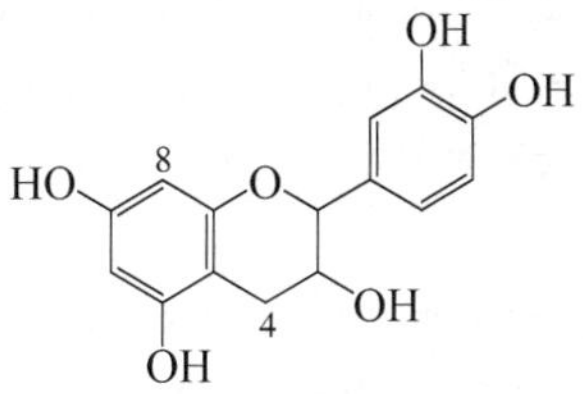

Catechin

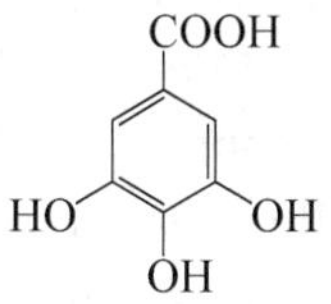

Gallic Acid

When tannins are extracted into hot water, the hydrolazable ones are partially hydrolyzed, meaning that free gallic acid is also found in tea. The tannins, by virtue of their phenolic groups, and gallic acid by virtue of its carboxyl groups, are both acidic. If calcium carbonate, as base, is added to tea water, the calcium salts of these acids are formed. Caffeine can be extracted from the basic tea solution with chloroform, but the calcium salts of gallic acid and the tannins are not chloroform soluble and remain behind in the aqueous solution.

The brown color of a tea solution is due to flavonoid pigments and chlorophylls, as well as their respective oxidation products. Although chlorophylls are somewhat chloroform soluble, most of the other substances in tea are not. Thus, the chloroform is easily removed by distillation (bp is 61℃) to leave the crude caffeine. The caffeine may be purified by recrystallization or by sublimaton.

In a second part of this experiment, caffeine will be converted to a derivative. A derivative of a compound is a second compound, of known melting point, formed from the original compound by a simple chemical reaction. In trying to make a positive identification of an organic compound, it is often customary to convert it into a derivative. If the first compound, caffeine in this case, and its derivative both have melting points which match those reported in the chemical literature (e.g., a handbook), it is assumed that there is no coincidence and that the identity of the first compound, caffeine, has been definitely established.

Caffeine is a base and will react with an acid to give a salt. Using salicylic acid, a derivative salt of caffeine, caffeine salicylate, will be made in order to establish the identity of the caffiene isolated from tea leaves.

Special instructions. Be careful when handling chloroform. It is a toxic solvent, and you should not breathe it excessively or spill it on yourself. When discarding spent tea leaves, do not put them in the sink because they will clog the drain. Dispose of them in a waste container.

Procedure

Place 25g of dry tea leaves, 25g of calcium carbonate powder, and 250ml of water in a 500ml three neck round bottom flask equipped with a condenser for reflux. Stopper the unused openings in the flask and heat the mixture under reflux for about 20 minutes. Use a Bunsen burner to heat. While the solution is still hot, filter it by gravity through a fluted filter using a fast filter paper such as E&D No. 617 or S&S No. 595. You may need to change the filter paper if it clogs.

Cool the filtrate to room temperature and, using a separatory funnel, extract it twice with 25ml portions of chloroform. Combine the two portions of chloroform in a 100ml round bottom flask. Assemble an apparatus for simple distillation and remove the chloroform by distillation. Use a steam bath to heat. The residue in the distillation flask contains the caffeine and is purified as described below (crystallization). Save the chloroform that was distilled. You will use some of it in the next step. The remainder should be placed in a collection container.

Crystallization (Purification)

Dissolve the residue obtained from the chloroform extraction of the tea solution in about 10ml of the chloroform that you saved from the distillation. It may be necessary to heat the mixture on steam bath. Transfer the solution to a 50ml beaker. Rinse the flask with an additional 5ml of chloroform and combine this in the beaker. Evaporate the now light-green solution to dryness by heating it on a steam bath in the hood.

The residue obtained on evaporation of the chloroform is next crystallized by the mixed solvent method. Dissolve it (which obtained on evaporation of the chloroform) in a small quantity (about 2 to 4ml) of hot benzene and add just enough highboiling (60℃ to 90℃) petroleum ether to turn the solution faintly cloudy. Alternatively, acetone may be used for simple crystallization without a second solvent. Cool the solution and collect the crystalline product by vacuum filtration using a Hirsch funnel. Crystallize the product the same way a second time if necessary, and allow the product to dry by allowing it to stand in the suction funnel for a while. Weigh the product. Calculate the weighty percentage yield based on tea and determine the melting point. If desired, the product may be further purified by sublimation as described in the next experiment.

The Derivative

Dissolve 0.20g of caffeine and 0.15g of salicylic acid in 15ml of benzene in a small beaker by warming the mixture on a steam bath. Add about 5ml of high boiling point (60℃ to 90℃) petroleum ether and allow the mixture to cool and crystallize. It may be necessary to cool the beaker in an ice water bath or to add a small amount of extra petroleum ether to induce crystallization. Collect the crystalline product by vacuum filtration using a Hirsch funnel. Dry the product by allowing it to stand in the air, and determine its melting point. Check the value against that in the literature. Submit the sample to the instructor in a labeled vial.

(Pavia Donald L. Introduction to Organic laboratory Techniques: a Contemporary Approach. W B Saunders Company, 1976.)

Words and Expression

caffeine   n. 咖啡因，咖啡碱
glucose   n. 葡萄糖

extract vt. 榨出，提取，萃取，蒸馏（出）；n. 萃取物，提取物
tannin n. 丹宁，丹宁酸，鞣酸
homogeneous adj. 均一的，均相的，均匀的，同质的
phenolic adj. 酚的
precipitate vt. 使沉淀；vi. 沉淀；n. 沉淀物
hydrolyze vi. 水解
gallic acid 五倍子酸
hydroxyl group 羟基
esterify v. （使）酯化
digalloyl group 鞣酰基
non-hydrolyzable adj. 不可水解的
catechin n. 儿茶酸
hydrolysable adj. 可水解的
carboxyl group 羧基
acidic adj. 酸的，酸性的
calcium carbonate 碳酸钙
chloroform n. 氯仿
flavonoid pigment 黄酮类颜料
chlorophyll n. 叶绿素
distillation n. 蒸馏
recrystallization n. 重结晶
sublimation n. 升华
derivative n. 衍生物；adj. 衍生的
salicylic acid 水杨酸
salicylate n. 水杨酸盐，水杨酸酯
three neck round bottom flask 三口烧瓶
condenser n. 冷凝器
stopper n. 塞子；v. 塞住
reflux n. 回流
Bursen burners 本生灯
filtrate n. 滤（出）液；v. 过滤
filter paper 滤纸
separatory funnel 分液漏斗
steam bath 蒸汽浴
distillation flask 蒸馏瓶
beaker n. 烧杯
rinse v. 冲洗，漂洗；n. 漂清，冲洗

petroleum ether   石油醚
ligroin   n. 轻石油，粗汽油
acetone   n. 丙酮
Hirsch funnel   赫氏漏斗
suction funnel   吸附漏斗
ice water bath   冰水浴
vial   n. 小瓶，小玻璃瓶；vt. 放……于小瓶中

Exercises

1. Answer the following questions
① Can you list several plants that contain caffeine?
② What kind of method can be used to isolate caffeine from tea?
③ How does man use caffeine in the daily life?

2. Put the following into Chinese

cellulose	glucose	chloroform	beaker
crystallization	purification	apparatus	filter paper
hydroxyl group	carboxyl group	benzene	acetone
evaporation	insoluble	condensation	residue

3. Put the following into English

蒸馏	重结晶	升华	过滤
酸	碱	盐	水杨酸
水杨酸钙	水杨酸甲酯	冷凝器	塞子
分液漏斗	水解	可水解的	三口烧瓶

Lesson 3   The Chemistry of Insulin

Insulin was isolated in crystalline form by Abel in 1926, and its chemical structure was elucidated by Sanger and his co-workers in the early 1950s. Sanger found that the insulin molecule is composed of two polypeptide chains, an A chain, consisting of 21 amino acid residues, and B chain, containing 30 residues. The two chains are connected by two disulfide bonds, and there is an additional disulfide linkage within the A chain.

The amino acid sequences of insulin for at least 28 species have been reported. Although most of these insulins are remarkably similar in amino acid composition and molecular weight, insulin from hagfish has been found to differ from human insulin in almost 50% of the residues. The two types of insulin of greatest importance to us, due to their therapeutic role, are those isolated from the pig and cow. Porcine insulin differs from human insulin in that it contains a C-terminal alanine in the B Chain, whereas human insulin has a *C*-terminal threonine. Because they are so similar in structure, it is possible to convert porcine insulin into human insulin. The sequential differences between bovine insulin (Figure 7) and human insulin are more striking in that, in addition to the porcine change, there are amino acid differences at positions 8 and 10 of the A chain. At position A-8, alanine replaces threonine, and at A-10, valine is substituted for isoleucine. Although the biologic activity is retained with these amino acid changes, it should be kept in mind that the more dissimilar the sequences, the greater is the potential for antigenicity. Predictably, bovine insulin is more antigenic than porcine insulin in humans.

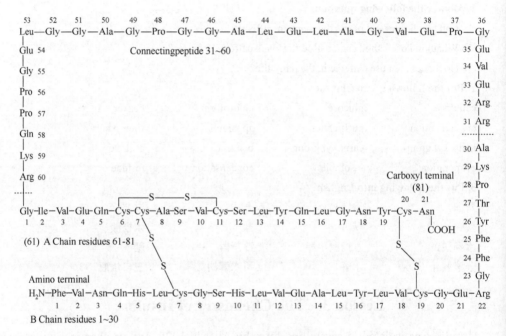

Figure 7　Bovine proinsulin

The three-dimensional arrangement of atoms in insulin has been determined by X-ray crystallography. From this work we can appreciate that insulin looks like a typical globular protein, despite its small size. Insulin crystallized at neutral pH is composed of six molecules of insulin organized into three identical dimers. The three dimers are arranged around two zinc ions, each of which is coordinated to the imidazole nitrogens of three B-10 histidines. Presumably, insulin is stored in granules in the β-cell as the hexamer, but the physiologically active form of insulin is the monomer.

The chemical synthesis of insulin was achieved independently by three groups during the mid-1960s. All three used the method of separate synthesis of the A and B chains followed by random combination. The process involved more than 200 steps and took several years to accomplish. Using the modern-day methods of solidstate synthesis developed by Merrifeld and Marglin, the same results have been obtained in just a few days.

The major difficulty encountered in the synthesis of insulin was the correct positioning of the three disulfide bonds. This was because of the many ways in which the sulfhydryl groups could combine. The problem of poor yields was partially resolved by treating the sulfhydryl form of one chain with the S-sulfonated derivative of the other. Proper positioning of the disulfide bonds has been further facilitated through conformational-directed disulfide bond formation. By reversible cross-linking of amino acids known to be in juxtaposition in the three-dimensional structure, the formation of the correct disulfide bonds can be enhanced.

Another novel approach to the synthesis of insulin has been to circumvent the problem of disulfide bond formation. The appropriate disulfide bond was first formed between fragments of the A and B chain. Then the fragments, with the disulfide bonds correctly positioned, were condensed in an order fashion so as to obtain the final product. Human insulin synthesized in this manner may be biologically equivalent to the natural hormone. Although the fragment condensation approach is a significant improvement over other methods for synthesizing proteins containing disulfide bonds, it is not likely to become a commercial source of insulin in the near future. Nature sources of insulin still appear to be sufficiently large and are cheaper to obtain.

Besides, concern over the future adequacy of insulin supplies may soon become academic. Advances in recombinant DNA technology have moved us closer to the synthesis of insulin by bacteria. Ullrich and co-workers have successfully incorporated the gene for rat insulin into bacterial plasmids; Gilbert's team from Harvard has reported the successful synthesis of rat proinsulin from another bacterial clone; and investigates from Genetech reportedly have bacterial strains that synthesize either the A chain or the B chain of insulin. Although none of these advances are commercially feasible at the present time, the stage has been set for the construction of bacterial plasmids containing human insulin genes. Such an achievement would assure an almost limitless supply of insulin for the world's 60 million diabetics.

The availability of insulin analogs and chemically and enzymatically modified insulin derivatives has provided a means of studying the relationship between the chemical structure and the biologic activity of the insulin molecule. Neither the A chain nor the B chain is active in the fully reduced form.

There is no apparent loss of activity in the insulin molecule when the *C*-terminal alanine is removed from the B chain of porcine or bovine insulin, but the biologic activity is diminished appreciably when both the *C*-terminal alanine and the *C*-terminal asparagine are removed. The removal of the second and third amino acids from the *C*-terminal of the B-chain decreases the potency slightly, and desoctapeptide insulin has almost no activity and retains little ability to dimerize. Removing or chemically modifying the *N*-terminal glycine of the A chain substantially decreases biologic activity, but activity is retained if the B-chain *N*-terminal phenylalanine is deleted. Chemical modification of the side carboxyl groups or the tyrosine residues also leads to inactivity.

The information obtained from studies such as these has confirmed that the conformation of the insulin molecule is critical to its activity, and has provided a basis for Pellin and collaborators to propose a receptor-binding region within the insulin molecule.

(Foye William O. Principles of Medicinal Chemistry. London: Hery Kenry Kimpton Publishers, 1981.)

Lesson 4   Tablets (The Pharmaceutical Tablets Dosage Form)

Role in therapy

The oral route of drug administration is the most important method of administering drugs for systemic effects. Except in cases of insulin therapy, the parenteral route is not routinely used for self-administration of medication. The topical route of administration has only recently been employed to deliver drugs to the body for systemic effects, with two classes of marked products: nitroglycerin for the treatment of angina and scopolamine for the treatment of motion sickness. Other drugs are certain to follow, but the topical route of administration is limited in its ability to allow effective drug absorption for systemic drug action. The parenteral route of administration is important in treating medical emergencies in which a subject is comatose or cannot swallow, and in providing various types of maintenance therapy for hospitalized patients. Nevertheless, it is probable that at least 90% of all drugs used to produce systemic effects are administered by the oral route. When a new drug is discovered, one of the first questions a pharmaceutical company asks is whether or not drug can be effectively administered for its intended effect by the oral route. If it can not, the drug is primarily relegated to administration in a hospital setting or physician's office. If patient self-administration cannot be achieved, the sales of the drug constitute only a small fraction of what the market would be otherwise. Of drugs that are administered orally, solid oral dosage forms represent the preferred class of product. The reasons for this preference are as follows. Tablets and capsules represent unit dosage forms in which one usual dose of the drug has been accurately placed. By comparison, liquid oral dosage forms, such as syrups, suspensions, emulsions, solutions, and elixirs, are usually designed to contain one dose of medication in 5 to 30ml. The patient is then asked to measure his or her own medication using a teaspoon, tablespoon, or other measuring devices. Such dosage measurements are typically in error by a factor ranging from 20% to 50% when the drug is self-administered by the patient.

Liquid oral dosage forms have other disadvantages and limitations when compared with tablets. They are more expensive to ship(one liquid dosage weighs 5g or more versus 0.25 to 0.4g for the average tablet), and breakage or leakage during shipment is a more serious problem with liquids than with tablets. Taste masking of the drug is often a problem (if the drug is in solution even partially). In addition, liquids are less portable and require much more space per number of doses on the pharmacist's shelf. Drugs are in general less stable (both chemically and physically) in liquid form than in a dry state and expiration dates tend to be shorter. Careful attention is required to assure that the product will not allow a heavy microbiologic burden to develop on standing or under normal conditions of use once opened (preservation requirements). There are basically three reasons for having liquid dosage forms of a drug: ① The liquid form is what the public has come to expect for certain types of products (e.g., cough medicines). ② The product is more effective in a liquid form (e.g., many adsorbents and antacids). ③ The drugs are used fairly commonly by young children or the elderly, who have trouble swallowing the solid oral dosage forms.

Properties

The objective of the design and manufacture of the compressed tablet is to deliver orally the correct amount of drug in the proper form at or over the proper time and in the desired location, and to have its chemical integrity protected to that point. Aside from the physical and chemical properties of the medicinal agent(s) to be formulated into a tablet, the actual physical design, manufacturing process, and complete chemical makeup of the tablet can have a profound effect on the efficacy of the drug(s) being administered.

A tablet (i) should be an elegant product having its own identity while being free of defects such as chips, cracks, discoloration, contamination, and the like; (ii) should have the strength to withstand the rigors of mechanical shocks encountered in its production, packaging, shipping, and dispensing; and (iii) should have the chemical and physical stability to maintain its physical attribute over time. Pharmaceutical scientists now understand that various physical properties of tablets can undergo change under environmental or stress conditions, and that physical stability, through its effect on bioavailability in particular, can be of more significance and concern in some tablet systems than chemical stability.

On the other hand, the tablet (i) must be able to release the medicinal agents in the body in a predictable and reproducible manner and (ii) must have a suitable chemical stability over time so as not to allow alteration of the medicinal agent(s). In many instances, these sets of objectives are competing. The design of tablet that emphasizes only the desired medicine effects may produce a physically inadequate product. The design of a tablet emphasizing only the physical aspects may produce tablets of limited and varying therapeutic effects. As one example of this point, Meyer and associates present information on 14 nitrofurantoin products, all of which passed the compendia physical requirements, but showed statistically significant bioavailability differences.

(Lachman Leon, et al. The Theory and Practice of Industrial Pharmacy. 3rd ed.
Lea and Febiger, Philadelphia, 1986.)

Words and Expressions

therapy　n. 治疗
administration　n. 管理，（药的）服法
parenteral　adj. 非肠道的
topical　adj. （医）局部的
nitoglycerin　n. 硝酸甘油，三硝酸甘油酯
angina　n. （医）心绞痛
scopolamine　n. 东莨菪碱
comatose　adj. 昏迷的，麻木的
relegate　vt. 驱逐的，使降级，把……归类
tablet　n. 药片，片剂
capsules　n. 胶囊，囊
dosage　n. 配药，剂量，用量，一剂，一服，一次分发量
syrup　n. 糖浆
suspension　n. 悬浮液
emulsion　n. 乳浊液，乳剂
elixir　n. （药）甘香酒剂
breakage　n. 破损
taste masking　n. 味觉模糊
expiration　n. 呼气，吐气
microbiologic　adj. 微生物学的
preservation　n. 保存，储藏，防腐
rigor　n. 僵硬
dispense　vt. 分配，调剂，配药
bioavailability　n. 生物利用度
nitrofurantoin　n. 呋喃妥因

Exercises

1. Put the following into Chinese

absorption	action	treat	medication
medicine	pharmaceutical	compress	quality
quantity	uniformity	measure	composite

2. Put the following into English

片剂	胶囊	糖浆	悬浮剂
乳剂	溶液	酊剂	丸剂
稀释剂	黏合剂	崩解剂	润滑剂
香味剂	甜味剂		

Lesson 5 Sterile Products

Sterile products are dosage forms of therapeutic agents that are free of viable microorganisms. Principally, these include parenteral, ophthalmic, and irrigating preparations. Of these, parenteral products are unique among dosage forms of drugs because they are injected through the skin or mucous membranes into internal body compartment. Thus, because they have circumvented the highly efficient first line of body defense, the skin and mucous membranes, they must be free from microbial contamination and from toxic components as well as posses an exceptionally high level of purity. All components and processes involved in the preparation of these products must be selected and designed to eliminate, as much as possible, contamination of all types, whether of physical, chemical, or microbiologic origin.

Preparations for the eye, though not introduces into internal body cavities, are placed in contact with tissues that are very sensitive to contamination. Therefore, similar standards are required for ophthalmic preparations.

Irrigation solutions are now also required to meet the same standards as parenteral solutions because during an irrigation procedure, substantial amounts of these solutions can enter the bloodstream directly through open blood vessels of wounds or abraded mucous membranes. Therefore, the characteristics and standards presented in this chapter for the production of large-volume parenteral solutions apply equally to irrigation solutions.

Sterile products are most frequently solutions or suspensions, but may even be solid pellets for tissue implantation. The control of a process to minimize contamination for a small quantity of such a product can be achieved with relative ease. As the quantity of product increases, the problems of controlling the process to prevent contamination multiply. Therefore, the preparation of sterile products has become a highly specialized area in pharmaceutical processing. The standards established, the attitude of personnel, and the process control must be of a superior level.

Vehicles

By far the most frequently employed vehicle for sterile products is water, since it is the vehicles for all natural body fluids. The superior quality required for such use is described in the monograph on *Water for Injection in the USP*. Requirements may be even more stringent for some products, however.

One of the most inclusive tests for the quality of water is the total solids content, a gravimetric evaluation of the dissociated and undissociated organic and inorganic substances present in the water. However, a less time-consuming test, the electrolytic measurement of conductivity of the water, is the one most frequently used. Instantaneous measurements can be obtained by immersing electrodes in the water and measuring the specific conductance, a measurement that depends on the ionic content of the water. The conductance may be expressed by the meter scale as conductivity in micromhos, resistance in megohms, or ionic content as parts per million (ppm) of sodium chloride. The validity of this measurement as an indication of the purity of the water is inferential in that methods of producing high-purity water, such as distillation and reverse osmosis, can be expected to remove undissociated substances along with those that are dissociated. Undissociated substances such as pyrogens, however, could be present in the absence of ions and not be disclosed by the test.

Additional tests for quality of Water for Injection with permitted limits are described in the USP monographs. When comparing the total solids permitted for Water for Injection with that for Sterile Water for Injection, one will note that considerably higher values are permitted for Sterile Water for Injection. This is necessary because the latter product has been sterilized, usually by a thermal method, in a container that has dissolved to some extent in the water. Therefore, the solids content will be greater than for the nonsterilized product. On the other hand, the 10 ppm total solids officially permitted for Water for Injection may be much too high when used as the vehicle for many products. In practice, Water for Injection normally should not have a conductivity of more than 1 micromho (1 megohm, approximately 0.1 ppm NaCl).

Added substances

Substances added to a product to enhance its stability are essential for almost every product. Such substances include solubilizers, antioxidants, chelating agents, buffers, tonicity contributors, antibacterial agents, antifungal agents, hydrolysis inhibitors, antifoaming agents, and numerous other substances for specialized purposes. At the same time, these agents must be prevented from adversely affecting the product. In general, added substances must be nontoxic in the quantity administered to the patient. They should not interfere with the therapeutic efficacy nor with the assay of the active therapeutic compound. They must also be present and active when needed throughout the useful life of the product. Therefore, these agents must be selected with great care, and they must be evaluated as to their effect upon the entire formulation. An extensive review of excipients used in parenteral products and the means for adjusting pH of these products has recently been published and should be referred to for more detailed information.

Formulation

The formulation of a parenteral product involves the combination of one or more ingredients with a medicinal agent to enhance the convenience, acceptability, or effectieness of the product. Rarely is it preferable to dispense a drug singly as a sterile dry powder unless the formulation of a stable liquid preparation is not possible.

On the other hand, a therapeutic agent is a chemical compound subject to the physical and chemical reactions characteristic of the class of compounds to which it belongs. Therefore, a careful evaluation must be made of every combination of two or more ingredients to ascertain whether or not adverse interactions occur, and if they do, of ways to modify the formulation so that the reactions are eliminated or minimized.

Production

The production process includes all of the steps from the accumulation and combining of the ingredients of the formula to the enclosing of the product in the individual container for distribution. Intimately associated with these processes are the personnel who carry them out and the facilities in which they are performed. The most ideally planned processes can be rendered ineffective by personnel who do not have the right attitude or training, or by facilities that do not provide an efficiently controlled environment.

To enhance the assurance of successful manufacturing operation, all process steps must be carefully reduced to writing after being shown to be effective. These written process steps are often called standard operating procedures (SOPs). No extemporaneous changes are permitted to be made in these procedures; any change must go through the same approval steps as the original written SOPs. Further, extensive records must be kept to give assurance at the end of the production process that all been performed as prescribed, an aspect emphasized in the FDA's Good Manufacturing Practices. Such in-process control is essential to assuring the quality of the product, since these assurances are even more significant than those from product release testing. The production of a quality product is a result of the continuous, dedicated effort of the quality assurance, production, and quality control personnel within the plant in developing, performing, and confirming effective SOPs.

(Lachman Leon, et al. The Theory and Practice of Industrial Pharmacy.
3rd ed, Lea and Febiger, Philadelphia, 1986.)

Exercises

1. Put the following into Chinese

parenteral   ophthalmic   irrigating   microorganisms
contamination specialize   conductivity   pyrogens

2. Put the following into English

灭菌产品   反相渗透   蒸馏   测量仪
电极   电导率   微生物   组织

Lesson 6 Natural Products

The naturally occurring organic compounds produced by living organisms have fascinated human beings for centuries. Indeed, traditional folk medicines from many cultures are effective because individual organic compounds that occur in the native plants exhibit biological activities that may be used to treat various maladies. Compounds of nature have been widely used in modern medicine as analgesics to relieve pain and as antibacterial, antiviral, anticancer, and cholesterol-lowering agents to treat or cure numerous ailments and diseases. On the other hand, some plants produce compounds that are toxic and may even cause death. For example, Socrates drank a fatal potion prepared from an extract of hemlock. Other plants have attracted attention because of the flavors they possessed or the odors they released. Organic compounds found in these plants are often incorporated as additives in the food and perfume industries. Some natural products such as those found in flowering plants are highly colored, and these may be used to prepare the dyes that color paints and fabrics. However, natural products have not only drawn attention because of their practical applications in our daily lives; they are also important for the various scientific challenges they present. Such challenges include understanding the basis of their sensory and biological properties and developing means for their chemical synthesis. Indeed, natural products provided chemists with many interesting experimental problems during the period when chemistry was emerging from alchemy into a more exact science. Today the isolation, identification, and synthesis of natural products, especially those having interesting and useful biological activity, remain important areas of research for modern organic chemists.

Historically, most natural products were extracted from plants rather than from animals, but more recently microorganisms are assuming increased importance as sources of natural products, particularly since the advent of molecular biology. The isolation of a natural product in pure form normally represents a significant challenge, mainly because even the simplest plants and microorganisms contain a multitude of organic compounds, often in only minute amounts. For example, a kilogram of plant material may only yield milligram quantities, or even less, of the desired natural product. The general approach used to isolate and purify a natural product is labor-intensive and typically starts with grinding a plant or other organism into fine particles. The resulting material is then extracted with a solvent or a mixture of solvents. Volatile natural products in the extract can be detected and isolated by gas chromatography. However, many natural products are relatively nonvolatile, and removing the solvent used for the extraction yields an oil or gum that requires further separation by chromatographic methods to obtain the various components in pure form. Often a series of chromatographic separation are required before the individual compounds are pure.

The next step is to determine the structure of the isolated products. Traditional approaches to structural elucidation involved performing standard qualitative tests for the various functional groups to identify those present in the molecule. The compound was then subjected to a series of simple chemical degradations to obtain simpler compounds. If a known compound was formed as a result of these transformations, it was then possible to work backward and deduce the structure of the original natural substances. More recently, instrumental techniques including infrared and nuclear magnetic resonance spectroscopy and mass spectrometry have greatly facilitated the determination of structures. However, the most reliable means of determining the structure of an unknown substance is by X-ray diffraction. In this technique, a crystal of sample is irradiated with X-rays to provide a three-dimensional map of the positions of each atom within the molecule. Applying this method requires that the unknown itself, or a suitable derivative of it, should be a crystalline solid, but this is not always possible.

The final goal of organic chemists working in this exciting field is often the laboratory synthesis of the natural product. While the total synthesis of a natural product may represent mainly an intellectual challenge, more often the undertaking is a unique opportunity to develop and demonstrate the utility of new synthetic techniques in organic chemistry. In cases where the natural product has medicinal applications, the development of an efficient synthetic route may be of great importance, especially when the supply of the material from the natural source is severely limited or its isolation is difficult. Through synthesis it is often possible to prepare derivatives of the natural product that possess improved biological properties.

There are a number of classes of natural products, and a detailed discussion of these compounds may be found in your lecture textbook. Representative compounds from the different classes range from the simple, as in the case of formic acid, HCO_2H, the irritating ingredient in ant venom, to the extraordinarily complex, as engendered in DNA, for example. The example of reserpine [Figure 8(a)], taxol [Figure 8(b)], and calicheamicin [Figure 8(c)] are illustrative of the diversity and complexity of the molecules of nature that have inspired research efforts of contemporary organic chemists and tested their creativity.

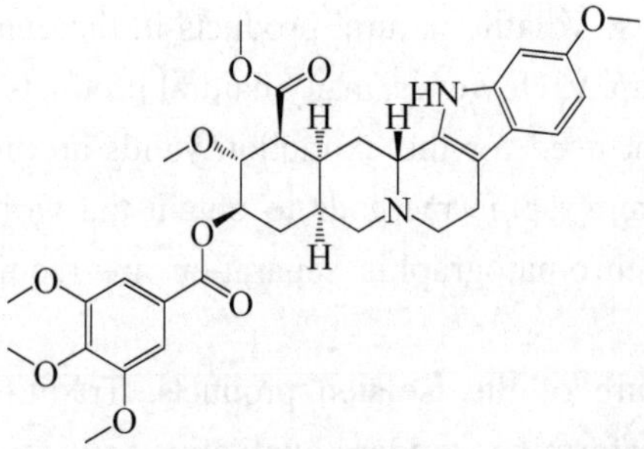

(a) reserpine

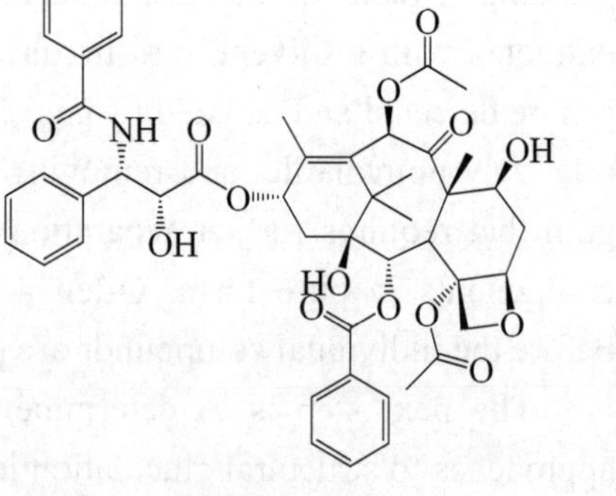

(b) taxol

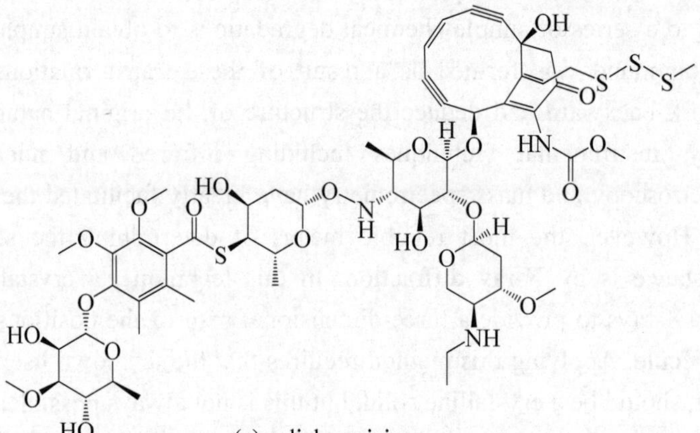

(c) calicheamicin

Figure 8　Reserpine(a), taxol(b) and calicheamicin(c)

Reserpine (a) is an alkaloid that was isolated from the Indian snakeroot Rauwolfia serpentine benth in 1952. Taxol (b) was isolated from the bark of the Pacific yew tree, Texus brevifolin, in 1962 as part of an extensive program directed toward the discovery of novel anticancer agents that was conducted jointly by the department of Agriculture and National Cancer Institute. Calicheamicin (c), is perhaps the most prominent member of recently discovered class of anticancer agents that contain an enediyne moiety, which consists of a double and two triple bonds.

Words and Expressions

antibacterial   adj. 抗菌的
adrenergic   adj. 肾上腺素的
antiviral   adj. 抗病毒的
anticancer   adj. 抗癌的
ailment   n. 疾病
hemlock   n. 毒芹，毒芹属植物；从毒芹提炼的毒药
postganglionic   adj. （神经）节后的
neuron   n. 神经细胞，神经元
ovarian   adj. 卵巢的
chemotherapeutic   adj. 化学疗法的
taxol   n. 紫杉醇
clinical   adj. 临床的

Lesson 7   Discovery of Sulfa Drugs

The sulfanilamides were the first antibacterial drugs invented by chemists, and a fascinating story underlies the discovery of the medicinal properties of these compounds. Although many scientists played important parts in the discovery, a key individual in the effort was Gerhard Domagk. Born in Lagos, Germany, in 1895, Domagk attended the University of Kiel intending to become a doctor, but World War Ⅰ interrupted his medical studied. Following the armistice in 1918, he reentered the University of Kiel and earned his medical degree in 1921. After a brief career in academia, he moved to I.G. Farbenindustrie (I.G.F.), the German dye cartel, where his responsibility was testing the pharmacological properties of the new dyes being synthesized by chemists.

At the time Domagk joined I.G.F., there were no antibacterials. This presented a serious health problem because bacteria were known to be the agents that cause pneumonia, meningitis, gonorrhea, and streptococcic and staphylococcic infections. The I.G.F. team of chemists and pharmacologists set out to find compounds that would kill these microbes without harming their animal or human hosts, and they formulated a plan to determine whether certain dyes might be bactericidal. Their strategy evolved from the observation that particular dyes, specifically those containing a sulfonamide group, seemed to be particularly "fast" or tightly bound, to wool fabrics, indicating their affinity for the protein molecules comprising wool. Because bacteria are proteinaceous in nature, the researchers reasoned that the dyes might fasten to the bacteria in such a way as to inhibit or kill them selectively. As we shall see, this simple hypothesis was partially correct: the sulfoamido group was indeed essential, but the part of the molecule that made the substance function as a dye was irrelevant to its effectiveness as a bactericide.

One dye Domagk tested on laboratory mice and rabbits infected with streptococci was called Prontosil. This compound, was found to be strongly disinfective against these bacteria and could be tolerated by animals in large doses with no ill effects. This discovery of the bactericidal effect of Prontosil in animals was probably made in early 1932; I.G.F. applied for a patent in December of that year.

Clinical tests on human patients apparently began soon after this, but record is confused. Some accounts say that before any other tests had been made on humans, Domagk gave a dose of Prontosil in desperation to his deathly ill young daughter, who had developed a serious streptococcal infection following a needle prick; the girl then made a rapid recovery. Others report that the first clinical test was on a ten-month-old boy who was dying staphylococcal septicemia. His doctor, R. Forster, was a friend of Domagk's superior at I.G.F. and through him learned about a red dye (Prontosil) that was miraculously effective in animals against streptococci. Since the baby was nothing to lose if the dye was not effective against staphylococci, Forster gave the child two dose of the red dye; complete recovery rapidly followed.

Regardless of which of these two stories is correct, or whether both are true, it was widely recognized by the middle 1930s that the discovery of the bactericidal properties of Prontosil was a medical miracle, for which Domagk was awarded the Nobel Prize in Physiology or Medicine in 1939. There were other important developments in the years between 1933 and 1939, however. Domagk did not publish the results of his tests of Prontosil on animal infections until February 1935, more than two years after the work was done. Learning of his results, the Trefouels, a wife and husband team at the Pasteur Institute in Paris, were prompted to test the bactericidal properties of several compounds, all of which were "azo" dyes closely related in chemical structure to Prontosil. The feature common to their dyes was the sulfonamide portion, but the other parts of the molecules differed significantly. Remarkably, they found that the antibacterial properties of these dyes were virtually identical with those of Prontosil.

This finding led to an explanation of a puzzling fact about Prontosil: it was ineffective against bacteria in vitro but was strongly effective in vivo. Apparently, a metabolic process within animals was necessary to make the sulfonamide dyes antibacterial. The Trefouels reasoned that the dye is broken into two parts in animals and only the sulfonamide portion is effective as an antibacterial. To prove this, they synthesized the sulfonamide component of Prontosil, which was the known compound *p*-aminobenzenesulfonamide, or sulfanilamide, and found it to be as effective as Prontosil against bacterial infections. Comparison of the formulas of Prontosil and sulfanilamide makes it clear that cleavage of Prontosil at the azo double bond affords the skeleton of sulfanilamide. This cleavage occurs biochemically when Prontosil is injected or imbibed, and the sulfanilamide so produced is the actual antibacterial agent. The original hypothesis that sulfonamide dyes would be bactericidal was thus partly misconceived, in that only the sulfonamide part of the dye molecule kills microbes; the fact that it was a part of a dye molecule was incidental. Figure 9 is the structure of prontosil and sulfanilamide.

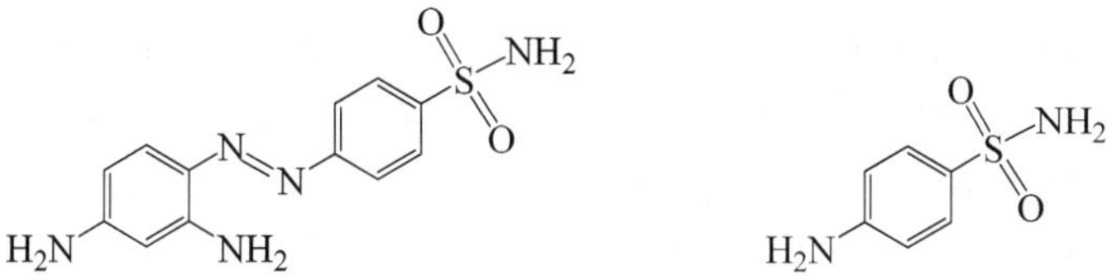

(a) prontosil (b) sulfanilamide

Figure 9 The structure of prontosil and sulfanilamide

Interestingly, the Trefouels' observations made the patent on Prontosil filed by I.G.F. useless. Sulfanilamide had been synthesized and patented many years before as a dye intermediate, but the patent had expired by the time the substance was found to be a potent bactericide. Moreover, their findings led to clinical trials of sulfanilamide in France, England, and the United States, all of which were highly successful. One case that gave great publicity to the new drug was the use of Prontosil to save the life of Franklin D. Roosevelt Jr., son of the president. In 1936, young Roosevelt was dying from a treptococcic infection when his mother convinced a doctor to administer Prontosil, which saved his life.

By 1947 over 5000 sulfonamides related to sulfanilamide had been prepared and tested for their efficacy as antibacterials. Although not all were effective, some were found to be better than sulfanilamide against certain infections. Of the thousands of compounds prepared and tested, the active ones are almost always those in which the only variation in structure is a change in the group of atoms attached to the nitrogen atom of the sulfonamide moiety.

Words and Expressions

pharmacological   adj. 药理学的
Prontosil   n. 百浪多息（一种磺胺类药物的商标名）
sulfanilamide   n. 磺胺

References

[1] 赵萱，等. 科技英语翻译 [M]. 北京：外语教学与研究出版社，2006.
[2] 郭正行，等. 科技英汉汉英翻译技巧 [M]. 天津：天津大学出版社，1998.
[3] 夏喜玲. 科技英语翻译技法 [M]. 郑州：河南人民出版社，2007.
[4] 范武邱. 实用科技英语翻译讲评 [M]. 北京：外文出版社，2001.
[5] 郑福裕. 科技论文英文摘要编写指南 [M]. 北京：清华大学出版社，2003.
[6] 秦荻辉. 科技英语语法 [M]. 北京：外语教学与研究出版社，2007.
[7] 周春晖. 科技英语写作 [M]. 北京：化学工业出版社，2003.
[8] 张志明. 日常英语与科技英语的不同表达方式 [J]. 沈阳大学学报，1993,1: 72-75.
[9] 李美，李井岗，仲亚丽. 试论科技英语的特点 [J]. 科教导刊，2016, 6: 43-44.
[10] Wang Yuehui. Exploration in Features and Skills of Scientific English Translation [C]. 2nd International Conference on Education Technology and Information System, ICETIS 2014: 72-75.
[11] 仗义兵. 化学专业英语教学初探——词汇构造法 [J]. 上饶师专学报，2000, 20(3): 60-64.
[12] 张文广，王祖浩. 无机物质的英文命名法 [J]. 化学教育，2006, 8: 28-30.
[13] 黄微雅. 医药化工类专业英语词汇的特点及教学策略 [J]. 广州化工，2009, 37 (2)：210-212.
[14] 顾霞敏, 黄微雅, 许海丹. 医药化工类专业英语教学内容的改革 [J]. 广州化工，2011, 39 (21)：192-193.
[15] 魏高原. 化学专业基础英语知识（Ⅰ）[M]. 北京：北京大学出版社，2004.
[16] 张军. 材料专业英语译写教程 [M]. 北京：机械工业出版社，2001.
[17] 石春成. 化学专业英语教程 [M]. 西安：西北工业大学出版社，2007.
[18] 胡鸣，等. 化学工程与工艺专业英语 [M]. 北京：化学工业出版社，2005.
[19] 匡少平. 材料科学与工程专业英语 [M]. 北京：化学工业出版社，2006.
[20] 曹同玉. 高分子材料工程专业英语 [M]. 北京：化学工业出版社，2008.
[21] 吴达俊. 制药工程专业英语 [M]. 北京：化学工业出版社，2006.
[22] 周光明. 化学专业英语 [M]. 重庆：西南师范大学出版社，2006.
[23] 唐冬雁，张磊. 应用化学专业英语 [M]. 哈尔滨：哈尔滨工业大学出版社，1999.
[24] 李洪涛. 材料科学与工程专业英语 [M]. 哈尔滨：哈尔滨工业大学出版社，2001.
[25] 朱红军，吕志敏. 应用化学专业英语教程 [M]. 2版. 北京：化学工业出版社，2011.
[26] 张裕平，姚树文，龚文君. 化学化工专业英语 [M]. 2版. 北京：化学工业出版社，2014.
[27] 张耀君. 纳米材料基础（双语版）[M]. 北京：化学工业出版社，2015.